International Environmental Risk Management

International Environmental Risk Management

A Systems Approach

Second Edition

Robert A. Woellner, John Voorhees,
and Christopher L. Bell

CRC Press
Taylor & Francis Group
Boca Raton London New York

CRC Press is an imprint of the
Taylor & Francis Group, an **informa** business

Second edition published 2020
by CRC Press
6000 Broken Sound Parkway NW, Suite 300, Boca Raton, FL 33487-2742

and by CRC Press
2 Park Square, Milton Park, Abingdon, Oxon, OX14 4RN

© 2021 Taylor & Francis Group, LLC

First edition published by CRC Press 1997

CRC Press is an imprint of Taylor & Francis Group, LLC

ISBN: 978-1-138-05452-3 (hbk)
ISBN: 978-0-367-51801-1 (pbk)
ISBN: 978-1-315-16668-1 (ebk)

Typeset in Palatino
by Deanta Global Publishing Services, Chennai, India

This book is dedicated to the people and organizations who are working hard to build a sustainable society and economy.

Contents

Preface: Environmental Management Systems: A Tool for Success

Robert A. Woellner

> Risk: It is that four-letter word used extensively in today's society to discuss everything from the dangers of operating a vehicle to uncertainty in the economy to the perils of certain lifestyle choices. Risk is inherently personal in nature and varies greatly between individuals, organizations, and businesses. For this reason, risk can be a divisive topic. What one person or group deems as an acceptable or "safe" behaviour, another will judge as "dangerous" or too risky. The one thing that most of us can agree on is that risks are all around us, regardless of the setting.
>
> **(Dotson 2018)**

This opinion, voiced by G. Scott Dotson, CIH, aptly summarizes the basis for this book on international environmental risk management. This book is not about judging risk; it has, as its foundation, the fact that risk exists in all of our daily activities, and it is the well-thought-out management of risk that allows us to survive and prosper.

In the business world, global environmental standards like ISO 14000, developed by the International Organization for Standardization (ISO), are the functional equivalent of such risk philosophies as they address environmental concerns on a worldwide basis. These standards are now being employed by national and multinational businesses to manage and control environmental risks and liabilities, thereby producing better performance and reducing adverse impacts on our communities and the ecosystem. Long-lasting solutions to global environmental problems need the attention and dedication of the people causing impacts, both positive and negative, on our environment. This book shows how environmental problems are resolved by competent individuals, environmental management systems, and techniques to reduce environmental impacts and simultaneously create revenue growth.

Governments can hamper or support businesses as they strive to achieve environmental excellence. Let's be clear: environmental regulations do not characteristically improve a business's bottom line, but the best of well-thought-out regulations do create an incentive for good businesses to utilize their intellectual, institutional, and financial power to improve their environmental performance and reduce risk. Properly designed and implemented regulations can level the playing field by requiring that the less well-intentioned businesses do not have an economic or operational advantage over the better-intentioned businesses that wish to appropriately manage and improve their environmental performance.

While governmental regulations can be helpful to businesses, government intervention can also retard environmental progress and can waste resources that could be used to prevent pollution rather than control it. In the U.S. during the 1970s, 1980s, and 1990s, tough environmental regulations were routinely fought by industry and frequently produced disincentives for businesses to minimize their environmental impacts. Former Environmental Protection Agency (EPA) Administrator William K. Reilly has criticized the U.S. regulatory approach because it was unsuccessful in solving the greatest environmental problems:

> We've had Love Canal, Valley of the Drums, and the Exxon Valdez. With virtually every case of a new environmental crisis, there is a new legislative priority and a new budget allocation. That has created a mix of programs that doesn't respect the biggest risks to health and ecology.

> **(Schneider 1993a)**

By reacting to environmental crises without considering the full range of consequences of regulatory policy, we have made costly mistakes. The mistakes need to be rectified quickly, and scarce resources must be reallocated to find permanent solutions to environmental problems throughout the world.

In 1982, government officials discovered high concentrations of dioxin in dirt roads in Times Beach, Missouri. The federal government caused a near panic by evacuating and permanently relocating 2,240 residents at a cost of $37 million. A decade later, the federal official who ordered the evacuation admitted that he had made a mistake because he had overestimated the danger presented by dioxin (Schneider 1993a). In 1985, Congress passed a law requiring local and state governments to remove asbestos from public buildings at a cost of $15–20 billion. In 1990, the EPA then admitted that the removal of asbestos frequently increased rather than decreased the asbestos exposure for building occupants (Schneider, 1993a). These two examples are representative of the wasted financial resources that could have been used to permanently improve our environmental quality of life.

Governments can be expected to make mistakes, particularly involving the environment where we still have limited knowledge of the long-term effects of chemicals. Businesses, however, cannot afford to commit significant errors when it comes to environmental protection. By requiring perfection from businesses in complying with rigid pollution control standards, we have set ourselves up for failure. The message sent to businesses is that it is more important to be in current compliance than to search for ways to dramatically reduce environmental impacts. Business leaders need to work with the government to improve their performance and, at the same time, identify and improve regulations that waste resources so that funds can be reallocated to solve more pressing environmental problems. In Yorktown, Virginia, the Amoco Oil Corporation was required to spend $31 million to rebuild the refinery's wastewater treatment plant to prevent benzene, a toxic chemical, from evaporating into the air. A study jointly funded by Amoco and the EPA found that federal regulations did not cover the major source of benzene emissions at the marine terminal of the refinery, which could have been abated for $6 million (Schneider 1993b). By working together, Amoco and the EPA solved an environmental problem that regulations had missed.

The Ocean Dumping Act is an example of well-intentioned pollution control legislation that does not address the real environmental problem. In the blistering summer of 1988, health officials closed beaches in New York and New Jersey after discovering bloody syringes, infectious hospital wastes, and dead dolphins scattered on beaches. Horrified citizens clasping pictures of disgusting debris descended on Congress, which then acted swiftly. In the autumn of 1988, Congress unanimously passed the Ocean Dumping Act, which prohibited New York City from dropping its processed waste into the sea. However, the contaminated beaches were caused not by sea dumping, but rather by the ancient combined sewer overflows in the metropolitan area, causing raw sewage and other noxious wastes to be flushed into the sewers after heavy downpours. Congress's solution required New York to spend vast sums of money landfilling the processed waste, but that did not solve the problem. While one politician gleefully declared, "This is a turning point in human history," it was, in actuality, like the Amoco refinery, just another example of government's misidentification of the source of the environmental problem.

In 1990, the EPA's Science Advisory Board concluded that environmental laws "are more reflective of public perceptions of risks than of scientific understanding of risk" (Schneider 1993a). Cataclysmic events in the late 1970s and 1980s created extraordinary public awareness of environmental problems and a demand for government action. Incidents like the Bhopal tragedy, the Exxon Valdez oil spill, the Deepwater Horizon oil spill, and the discovery of toxic waste dumps produced tough legislative restrictions on new development, existing business operations, and the disposal of all wastes. These events captured the media's attention and molded the public's viewpoint that environmental risks are unacceptable in the modern age.

The government's pollution control strategies have also created many adverse collateral consequences. Major environmental legislation in the U.S. has created a huge industry for lawyers and consultants to spend corporate resources squabbling with the government over past disposal practices and conformance to strict regulations.

> Policymakers, business leaders, and environmentalists have focused on
> the static cost impacts of environmental regulations and have ignored

the more important offsetting productivity benefits from innovation.
The whole process has spawned an industry of litigators and consultants
that drain resources away from real solutions.

(Porter and van der Linde 1995a)

In the past three decades, governments have concluded that they do not
have the resources, the personnel, or the technical ability to solve all of the
global environmental problems. The European Environmental Agency has a
minuscule $63 million budget (Official Journal of the European Union 2018),
a large percentage of which is spent on outside consultants who monitor
environmental conditions in Europe. The European Union (EU) created this
agency with no sweeping investigative powers and without the responsibility to inquire into its members' internal affairs. Contrast this with the large
multinational companies that are currently spending billions on pollution
control measures each year. U.S. manufacturers spend more than $100 bil-
lion annually trying to comply with environmental regulations. Total worldwide expenditures for environment-related products and services are well in
excess of $500 billion, exceeding the car market and reaching half the size of
the information technology market.

It is corporations, not governments, that are the only organizations with
the financial and human resources, technology, global reach, and motivation to achieve sustainability in future generations (Hart 1997). These funds
can be used to control pollution, but more importantly, they should be used
to eliminate the causes of pollution, reengineer production processes, and
cut costs. Businesses must learn how to turn pollution control strategies into
environmental opportunities that become major sources of revenue growth.

Focusing on sustainability requires putting business strategies to a new
test. Taking the entire planet as the context in which they do business,
companies must ask whether they are part of the solution to social and
environmental problems or part of the problem.

(Hart 1997)

Swift action by businesses is needed now to avoid repeating the mistakes
of the past. There is already great competition for the scarce resources that
will be allocated to clean up the past problems, and businesses cannot afford
to make mistakes in the future that can add to the existing contamination on
the earth.

The disposal of toxic chemicals is only a small part of a much greater problem that is harder to portray on a television screen or politician's soundbite,
but far more likely to have an enduring, negative impact on the environment
and business, resulting in greater challenges in the 21st century. Anthony D.
Cortese, the president of Second Nature, a nonprofit organization established
to improve environmental education, argues that rapid population growth

presents a fundamental danger to our environment (Cortese 1994). In 1830, the population reached 1 billion; by 1950, it grew to 2.6 billion, and then doubled by 1988. According to estimates and projections from the United Nations, World Health Organization, and other bodies, the world population as of June 2020 is nearly 7.8 billion, which is roughly double the number of people who were alive in 1970 (Worldometers 2020). At the current growth rate, the world population grows by one million people every 4.56 days. That is a population growth rate of more than 1.5 million people per week (History .com 2020). As the world's population increases, approximately 36 billion tons of topsoil are lost each year, with deforestation and changes in land use worsening the problem significantly (Borrelli 2017). Of the 7.8 billion people that inhabit the world, approximately 736 million still live in conditions of abject poverty (World Bank 2018). Currently, 55% of all humans in the world live in a city; it is estimated that by 2050, that figure will be 68%. By 2030, it is expected that the planet will have 43 megacities, each the home of more than 10 million inhabitants and the majority of them in developing regions (United Nations 2018). Unless there are extraordinary measures taken immediately on how we deal with our environment, "the kind of world implied by those numbers is unthinkable" (Magretta 1997).

As the population has increased and business is responding to marketplace needs, so has industrial production that causes pollution and depletion of natural resources. World industrialization has increased 100 times in the past 100 years (Clark 1989). With increased production comes more industrial pollution and wastes piled on top of the remnants of unregulated disposal of past refuse. Consumers are using and discarding products in greater numbers than ever before. Nations throughout the world are being forced to change how they protect the environment based on the growing population's escalating environmental use. And they must.

Acid rain is a major problem in cities and countries where coal combustion is unregulated. No country in the world has solved its air and water pollution problems. All have stockpiles of toxic chemicals leaching in waste dumps. Equally threatening is the consumer's use of products, such as automobiles, that contribute to immense pollution throughout the world. According to the World Bank, there are more than one billion automobiles in the world, mostly concentrated in cities where they will double current levels of energy use, smog precursors, and emission of greenhouse gases (Hart 1997). The growing dependency on cars in U.S. cities like Los Angeles and Asian cities like Bangkok and Kuala Lumpur has created virtual gridlocks in the absence of adequate transportation planning and infrastructure improvements. Transportation consulting firm INRIX determined that traffic congestion cost $305 billion in 2017 in the U.S. alone, as automobiles and trucks idled on the highway, wasting fuel, losing time, having accidents, and creating other related wastes (Schneider 2018). Our increasing use of the automobile is just one of many factors that need to be addressed globally if we can hope to reduce our environmental impacts and create a healthier, more sustainable environment. The consequences of not doing so are likely severe. In the context of the U.S., for example, the U.S. Global Change Research Program's massive 2018 climate report concludes:

> In the absence of more significant global mitigation efforts, climate change is projected to impose substantial damages on the U.S. economy, human health, and the environment. Under scenarios with high emissions and limited or no adaptation, annual losses in some sectors are estimated to grow to hundreds of billions of dollars by the end of the century. It is very likely that some physical and ecological impacts will be irreversible for thousands of years, while others will be permanent.

(USGCRP 2018)

Innovative educational efforts begun by Paul Hawken and others are demonstrating how people can be taught to become better caretakers of the environment. Cortese argues that we must change the way we educate people on environmental issues to create more responsible stewards of the environment for generations to come. Cortese and Hawken are urging schools to develop environmental education awareness curricula to be added to all subjects taught in secondary and higher education. By making environmentalism second nature to future business and government leaders, we can begin to change how products are produced and used, how waste is created and discarded, and how people can voluntarily minimize their individual environmental impacts. Many believe that the root cause of our neglect of the environment for many years was that as a nation and as a world, we believed our individual and collective impacts would result in only imperceptible environmental changes. This anthropocentric world view caused people to believe that they were the most important biological species and thus should have dominion over all of nature. People assumed that natural resources were free and inexhaustible, wastes could be assimilated in the environment, and societies and continents could adapt to any adverse changes (Cortese 1994). Years of production of nuclear weapons during the Cold War increased the legacy of environmental neglect and degradation. In the U.S. the total bill just to clean up all the hazardous, chemical, and radioactive waste sites is approximately $1 trillion.

Political and social pressures are forcing businesses to pay extraordinary costs to do business and use the environment. The definition of business and development costs is rapidly expanding to include new costs that have never been borne by business. Governments throughout the world are considering green taxes on carbon dioxide and other forms of pollution. Singapore already charges people to drive in its metropolitan center and has imposed quota premiums for car buyers. The right to buy a car is auctioned off to the public and can cost up to $20,000 for a luxury car (Mcinerney and White 1995).

Industrial and commercial expansion, and even residential growth, are being regulated by more government authorities than ever before. Citizens and environmental activist groups are erecting barriers for development and forcing businesses to become more responsive to their environmental and quality-of-life agendas. Conflicts are becoming less civil between asset holders, who have property interests and seek freedom from regulatory restraints, and stakeholders, who have use and enjoyment interests and urge greater access and less environmental degradation. Pollution knows no geographic, cultural, social, or political boundaries; hence, the line between asset holders and stakeholders is not precise. Under the modern theory of environmental dispute resolution, every interested party wants a seat at the bargaining table with an equal vote and a meaningful opportunity to be heard. While global issues (global warming, overpopulation, and deforestation, to name a few) are becoming more pronounced and political, there are many competing and urgent local issues that are creating tensions and disagreements on how we should allocate scarce public funds to contribute to environmental solutions. The discourse is becoming increasingly volatile because the public is blaming governments and businesses equally for allowing adverse impacts to the environment. People are more intolerant of any form of pollution, believing that governments and businesses owe them the duty to make their lives free of risks posed by industrial development. Public censure, amplified by politicians and worldwide media channels, is becoming perhaps the largest operations problem faced by businesses today.

Business leaders know that pollution is a significant cost of daily business. Any type of waste being discarded into the environment "is a sign that resources have been used incompletely, inefficiently, or ineffectively" (Porter and van der Linde 1995a). Pollution often reveals defective product designs and manufacturing processes, inattentive management practices, and unresponsiveness to environmental issues. It is a red flag for poor quality (Mcinerney and White 1995).

Environmental risks are inherent in business operations and can become oppressive. Few businesses in the world are immune to the threat of massive legal and financial liability for environmental claims, coupled with the loss of corporate reputation. Any claim or lawsuit, regardless of its accuracy, can create disastrous consequences. In 1997, two environmental activists in London claimed in a leaflet that the McDonald's hamburger chain destroys rain forests, flouts environmental concerns with its packaging, sells dangerously unhealthy food, seduces children into unhealthy eating habits, exploits staff (particularly African Americans and women), and is responsible for torturing animals (Montalbano 1997). McDonald's responded with a libel suit that lasted 314 days, which at the time was the longest trial in British history. McDonald's spent $16 million to win the case and recovered a paltry $98,000 in damages. However, the company's reputation, suffered with the widespread attention the libel suit drew. The case was featured on a two-night, three-hour British television reconstruction of the case called "McLibel!," a book titled *"McLibel: Burger Culture on Trial,"* and a website accessed by 9 million people (Montalbano 1997). There are many other examples of companies whose reputations have been damaged because of actions taken by environmental activists to publicize negative environmental impacts.

The myriad of complex environmental laws and regulations now require responsible corporate officers to monitor operations more closely. Failure to comply with these new obligations can lead to disastrous financial consequences. Investor scrutiny directed toward corporate management regarding environmental matters has never been more intense.

In the U.S., following Caremark International Inc.'s guilty plea in a criminal case that cost the company $250 million in related criminal and civil fines, Caremark shareholders sued the company's Board of Directors. They alleged that the Board bore personal responsibility for inadequately supervising the conduct of lower-level employees. The Board, however, was found to have properly discharged its duty because it had issued compliance guides and manuals that were reviewed and revised annually by outside advisors to the company; developed standard forms that assured compliance with applicable regulations; appointed a senior official to monitor and assure compliance; adopted an internal review process to assure that supervisors implemented compliance policies and procedures; evaluated compliance programs by an independent third-party auditor; issued a corporate code of conduct and ethics manual for all employees; and established a toll-free, confidential hotline for use in reporting infractions (Rothenberg et al. 1997).

The Caremark decision was widely publicized because of the implications to corporate executives and managers, who must not only understand how environmental issues affect the operation of their businesses but also correctly monitor them before damage occurs. Failure to exercise due diligence in taking steps to prevent and detect noncompliance can mean enhanced penalties pursuant to the U.S. Sentencing Guidelines for Organizations, but it is far more likely that it will result in a loss of market share, civil liability, and, perhaps, business failure. Preventing violations has become an important consideration in developing and implementing an environmental risk management and risk reduction program. Smart businesses seek to avoid these pitfalls by creating innovative solutions to pollution problems, potentially increasing their market share in the process.

Two environmental authors concluded, after a comprehensive environmental impact study of 10 global businesses, that "[c]ompanies that pollute the least have the highest quality products and services" (Mcinerney and White 1995). Companies can expand their market share and lower costs by drawing the connection between waste, margins, and quality (Mcinerney and White 1995). Systems that were developed for pollution control need to be shifted to pollution prevention. "Pollution control means cleaning up waste after it has been created. Pollution prevention focuses on minimizing or eliminating waste before it is created. Pollution prevention strategies depend on continuous improvement efforts to reduce waste and energy use" (Hart 1997).

In a 2017 message to CEOs, Blackrock CEO Larry Fink said, "To prosper over time, every company must not only deliver financial performance, but also show how it makes a positive contribution to society." There is a clear message that companies big and small, public and private, stand to incur real competitive benefits by investing in their Environmental, Social, and Governance ("ESG") efforts.

As detailed in their 2017 Annual Sustainability Report, the private equity firm Graham Partners considers ESG issues core to its mission. Graham Partners' sustainability program considers the three pillars, sustainability's "triple bottom line" philosophy. They explain that their ESG program is good for our *society* (investing in companies making a positive global contribution), the *planet* (minimizing energy and other resource use and demand, and reducing waste), and *profit* (creating real dollar savings and mitigating risk for their portfolio companies, which drives value to their investors). As a result of their sustainability activities, in 2017 and 2018 Graham Partners was named a top-ranked ESG program from LGT Capital Partners, a Switzerland-based fund of funds manager that surveys and ranks global private equity firms. Graham Partners was also featured in a <u>private equity case study</u> developed by LGT in 2017. Since these surveys and case study were completed, the firm has continued to advance its ESG program with several additional achievements, most recently detailed in Graham Partners 2019 Sustainability Report.

Environmental regulations have historically overlooked source reduction and focused almost exclusively on pollution control. Many businesses responded by creating technologies that focus on achieving stringent standards without eliminating the causes of pollution. Recycling is one means that has been used to address the problem after it occurs. Recycling can be an expensive, labor-intensive way of dealing with waste that does not address the real problem of eliminating the source of the waste. Recycling is an important first step because it makes people more aware of the scope of our environmental problems and allows them to participate directly and meaningfully in a community-wide effort to mitigate waste. It is, however, a poor substitute for the elimination of waste in the production process.

Companies need to use Design for Environment as a tool for creating products that are easier to recover, reuse, and recycle, if necessary. Design for Environment examines, during the design phase, all of the effects that a product could have on the environment. This includes the natural resources and energy used to create the product, as well as the packaging used in shipping the product to the consumer. This life-cycle product analysis begins and ends outside the boundaries of a company's operations and examines how customers use and dispose of the products (Hart 1997). The same type of analysis can be conducted for service businesses to review their internal operations and create a comprehensive plan to minimize environmental impacts.

We are now emerging into a new era of environmental reason where governments are learning how to create incentives for companies to innovate and improve product performance, quality, and management. Innovation can reduce the use of power, natural resources, and production materials.

Businesses that take a hard look at their internal operations will notice opportunities for the development of effective environmental management systems. Those systems not only reduce liabilities and risks, because less pollution is produced, but also increase profits because all resources are used more efficiently and effectively, making informed consumers more likely to buy the product or service even at higher per unit prices.

Professors Porter and van der Linde have argued persuasively that businesses now need to focus on outcomes, not technologies, by using *systems approaches*. Businesses must convince governments to enact environmental regulations that promote real innovation—the kind that business can recognize and implement rather than litigate in the court systems. Regulations should encourage upstream solutions to environmental problems and employ market incentives, drawing attention to resource and production inefficiencies. When the regulatory process becomes more stable and predictable, it can provide more opportunities and incentives for business to go beyond compliance and create revenue growth. Porter and van der Linde urge that industry must participate in setting standards from the beginning by developing strong technical capabilities among regulators and minimizing the time and resources consumed in the regulatory process.

New regulations alone are insufficient to drive business people to make correct decisions on how to manage and ultimately reduce their environmental risks and liabilities. Management systems must also be employed to show businesses how to take advantage of all available opportunities to redesign internal operations to limit environmental impacts. Cortese advocates gentle persuasion, not preaching, to change the way we do business to make meaningful environmental progress (Cortese 1994). He contends that the three most important ways to change corporate behavior on environmental issues are (1) make the business leaders believe they will not fail at the new venture; (2) show them how other companies are doing the same thing; and (3) demonstrate that the new behavior will result in financial gain.

A *systems approach* using global environmental management systems coupled with risk reduction techniques can assist in changing corporate behavior. Sustainability can be achieved if we make continual improvements in our processes and services and eliminate environmentally degrading sources of pollution. The *systems approach* can create sustainability and, at the same time, allow businesses to identify and limit their impacts, prevent pollution, restore natural resources, and minimize environmental risks and liabilities. Managing and reducing risks are part of a global ideological shift that is changing the way that businesses have operated and is creating opportunities for future generations to live free of past encumbrances.

The arena of corporate risk management has traditionally been characterized by two types of efforts: those performed to control existing risks and those performed to reduce future risks. Managing environmental risks effectively requires a comprehensive *systems approach* that combines both skills. As an example, the ISO 14000 series of standards can serve businesses as a model for an integrated series of management systems that identify, control, and monitor environmental risks. Additional risk management strategies include the use of information technology, risk-based approaches to internal and external environmental communication, insurance and risk transfer methodologies, and legal-risk reduction strategies.

Reducing future risks has as its central focus the identification of the causes of potential regulatory, public, and employee conflicts coupled with the elimination of adverse environmental impacts. Environmental management systems need to be augmented with auditing, litigation avoidance, collaborative decision making, alternative dispute resolution, involvement with proactive federal and state environmental programs, and other innovative risk reduction techniques. All of these areas need to be considered when creating or enhancing a comprehensive environmental risk management system.

The *systems approach* takes these risk management and reduction strategies and techniques and creates opportunities for further process innovations and solutions to environmental problems that will make the world a healthier, safer, and more prosperous place in which to live. It employs methods to avoid repeating past mistakes while at the same time encouraging future revenue growth. It provides strategies for corporate managers to follow when developing environmental risk management systems together with specific examples of techniques to eliminate risk. It contains guidelines for implementing environmental management systems and using daily experience with the system to enhance its effectiveness.

When environmental risk management systems are effectively implemented, they reduce pollution, minimize regulatory problems such as potential civil and criminal environmental liability, and maximize the public's and consumers' confidence in public safety and the businesses' efforts to preserve and protect the environment. The *systems approach* serves as a written guide for management and employees to conduct business operations in a safe, environmentally responsible manner. Most importantly, it demonstrates to the public that businesses are consciously and proactively managing their environmental concerns and are genuinely committed to safeguarding the environment. As eloquently summarized by Paul Hawken, "Good management is the art of making problems so interesting and their solutions so constructive that everyone wants to get to work and deal with them."

References

Borrelli, P., et al. 2017. An assessment of the global impact of 21st century land use change on soil erosion. *Nature Communications* 8(1).

Clark, W. 1989. Managing planet Earth. *Scientific American* 269.

Cortese, A.D. 1994. Earth Day 1995: Partnerships for sustainabilities. Presented at New England Earth Day Organizing Conference, John F. Kennedy School of Government. November 5. (On file with the authors.)

Doran, J., and G. Ryan. 2011. Regulation and firm perception, eco-innovation and firm performance. https://cora.ucc.ie/bitstream/handle/10468/781/Doran _and_Ryan_%282011%29.pdf?sequence=1&isAllowed=y.

Dotson, G.S. 2018. Beyond risk assessment: Integrating the risk sciences into the profession of industrial hygiene. *The Synergist* (September 21). https://synergist.ai ha.org/201809-beyond-risk-assessment.

Hart, S.L. 1997. Beyond Greening: Strategies for a sustainable world. *Harvard Business Review* (January/February).

Hart, D.M. 2018a. A better approach at the EPA: Use regulation to drive innovation. *Morning Consult* (August 3). https://morningconsult.com/opinions/better-ap proach-epa-use-regulation-drive-innovation/.

Hart, D.M. 2018b. When does environmental regulation stimulate technological innovation? *Information Technology & Innovation Foundation* (July 23). http://www 2.itif.org/2018-environmental-regulation-innovation.pdf?_ga=2.75340059.8075 716.1546550082-1741874278.1546550082.

Hawken, P., and W. McDonough. 1994. Seven steps to doing good business. *Inc.* (November).

International Organisation of Motor Vehicle Manufacturers (OICA). 2015. World vehicles in use – all vehicles. http://www.oica.net/wp-content/uploads// Total_in-use-All-Vehicles.pdf.

Magretta, J. 1997. Growth through global sustainability. *Harvard Business Review* (January/February).

Mcinerney, F., and S. White. 1995. *The total quality corporation.* New York: Truman Talley Books/Plume.

Montalbano, W.D. 1997. Verdict near in McLibel food fight. *Denver Post* (June 16).

Porter, M.E., and C. van der Linde. 1995a. Green and competitive: Ending the stalemate. *Harvard Business Review* (September/October).

Porter, M.E., and C. van der Linde. 1995b. Toward a new conception of the environment-competitiveness relationship. *Journal of Economic Perspectives* (Fall).

Rothenberg, E.B., C.A. Smith, and I.L. Feitshans. 1997. Caremark International, Inc. - Directors' obligation to assure compliance with governmental regulations. *Metropolitan Corporate Counsel* (April).

Schneider, B. 2018. Traffic's mind-boggling economic toll. *Citylab* (February 7). https://www.citylab.com/transportation/2018/02/traffics-mind-boggling-economic-toll/552488/.

Schneider, K. 1993a. New view calls environmental policy misguided. *New York Times* (March 21).

Schneider, K. 1993b. Unbending regulations incite move to alter pollution laws. *New York Times* (November 29).

United Nations. 2018. 68% of the world population to live in urban areas by 2050, says UN [Press release]. (May 16). https://www.un.org/development/desa/en/ news/population/2018-revision-of-world-urbanization-prospects.html.

U.S. Global Change Research Program (USGCRP). 2018. Chapter 29, Mitigation: The risks of inaction. In *Fourth National Climate Assessment, Volume II* (November 23). https://nca2018.globalchange.gov/.

World Bank. 2018. Decline of global extreme poverty continues but has slowed [Press release]. (September 19). http://www.worldbank.org/en/news/press-release/2018/09/19/decline-of-global-extreme-poverty-continues-but-has-slowed-world-bank.

Worldometers. 2020. World population. https://www.worldometers.info/ (accessed May 7).

Acknowledgments

We deeply appreciate the efforts of our friends and colleagues who contributed immensely to this book. We are very fortunate to have access to such strong leaders as Donald C. Graham, Dr. Richard C. Woellner, John C. Kolojeski, and Don Burklew. We are among the many beneficiaries of such visionaries as Samuel A. Graham, Sybil Fleming Graham, and Dr. Bill Brown. We thank the many supportive colleagues who assisted with this revision, especially Beth Boatright and Margaux Woellner. Finally, and most importantly, we would like to thank our wives Terri-Jo Woellner, Kim E. Voorhees, and Patricia A. Bell for all of their support, patience, and guidance.

Author Biographies

Robert A. Woellner, M.S., is the President of QUEST Environmental and QUEST Management International. Bob has more than 25 years of environmental risk assessment, industrial hygiene, indoor air quality, scientific investigative, and environmental management systems consulting experience. He serves on several boards, teaches college classes, and has published many articles and contributed to several books regarding environmental risk management. He has qualified and served in numerous jurisdictions as an expert in industrial hygiene, environmental monitoring, microflora/mold, meth, indoor air quality, environmental sampling, and risk management, for dozens of environmental contamination cases. He holds a Bachelor's degree in Geology from Middlebury College and a Master of Science degree in Marine Geology and Geophysics from the Rosensteil School of Marine & Atmospheric Science.

John Voorhees, J.D., a Shareholder at Greenberg Traurig, LLC, has practiced law for over 40 years and is a member of the American, Colorado, and Denver Bar Associations. He focuses his practice on compliance counseling, particularly on environmental, health care, fraud, and white collar matters. He has defended Superfund and Clean Water Act cases and a wide variety of other civil and criminal cases. He has represented hospitals and a disease management company in disputes with the Centers for Medicare and Medicaid Services (CMS) and is experienced at litigating hospitals' Medicare reimbursement claims against CMS, as well as fraud, waste, and abuse claims. He spent 11 years as a federal prosecutor in the Organized Crime & Racketeering Section of the U.S. Department of Justice before entering private practice in 1988. John received his law degree from the Columbus School of Law, Catholic University of America. He has represented QUEST since 1995.

Christopher L. Bell, J.D., is a partner in the international law firm of Greenberg Traurig LLP. He counsels clients and defends in civil and criminal enforcement cases on a wide range of issues, including compliance with water, waste, air, chemical, transportation, pollution prevention, cleanup, product stewardship, and safety requirements as well as international issues such as the EU's ROHS Directive and REACH Regulation, and issues such as climate change, sustainable development, nanotechnology, and social responsibility. He participates in legislative and regulatory advocacy involving most of the major environmental laws. He has extensive experience designing, implementing, and auditing environmental management and compliance assurance systems, including systems based on ISO 14001, RC

14001, and the Sentencing Guidelines. He was a lead U.S. international negotiator on ISO 14001 for a decade and also served as a negotiator in the development of the ISO 37001:2016 standard on antibribery management systems.
He is currently serving as an independent monitor overseeing compliance
with complex federal debarment and probation agreements arising from the
resolution of a major criminal environmental case.

This book has been a collaborative effort, with Messrs. Woellner taking
primary responsibility for Chapters 1–5, Bell Chapters 6–8, and Voorhees
Chapters 9–12. Although we frequently discuss the ISO 14000 standards, the
authors specifically decided to not focus our book on ISO 14000, but the environmental management topic more broadly. Interested readers are advised
to obtain and become familiar with the ISO 14000 standards by purchasing
the standards themselves, as well as reading the many excellent books and
articles on ISO 14000. The opinions expressed in this book are not necessarily the views of clients of Greenberg Traurig, QUEST Environmental, or
QUEST Management International.

1

The Context for Managing Environmental Risk

Robert A. Woellner

It is no coincidence that over 100 countries have joined together to create international standards for their industries and businesses to effectively manage their environmental impacts. The trend reflects a desire for systematic solutions to environmental problems. Global environmental management standards are addressed in the first five chapters, in which we provide the context for minimizing environmental liabilities by proactively managing environmental activities. The next three chapters are devoted to a more specific discussion on environmental risk management systems, and compliance and ethics programs, which have been developed to reduce environmental risks. The last four chapters focus on reducing litigation risks and costs and solving environmental problems without litigation. The development of global standards is the appropriate place to begin to explore how responsible solutions are created to address environmental concerns worldwide.

In 1946, the International Organization for Standardization (ISO) was founded as a worldwide federation to promote the development of international manufacturing, trade, and communication standards, thereby facilitating the international exchange of goods and services (Hall, 1996, and Murphy & Yates, 2009). ISO is a private sector, international standards body based in Geneva, Switzerland. It reviews input from government, industry, and other interested parties before it develops a standard. Understanding ISO and its mission and techniques allows for an appreciation of why the 9000 series for total quality management has been so popular and why the 14000 series portends even greater acceptance throughout the world.

1.1 ISO's Role and Mission

ISO's mission to develop manufacturing, trade, and communications standards assisted in the rebuilding of Europe after the Second World War. As

ISO grew rapidly, its purview became global and quickly involved nations outside Europe. The members of ISO are the standards organizations of different nations, and current membership stands at more than 160 countries. The U.S. is a full voting member of ISO, officially represented by the American National Standards Institute (ANSI). ISO strives to systematically develop globally accepted standards that are voluntarily adopted by the international business community. Specific application methods and techniques are not specified but left to the discretion of industry and government experts from around the globe.

The mission of ISO has two parts:

- To promote the development of standardization and related activities in the world to facilitate the international exchange of goods and services
- To develop cooperation in the sphere of intellectual, scientific, technological, and economic activity

Initially, this mission focused on technical performance specifications for products and standardized test methods. Currently, more than 320 technical committees are dedicated to the continual development of these types of standards. In 1979, however, a number of worldwide market trends led to a change in focus for ISO. These included the growth of industry throughout the world led by the post-Second World War boom in the U.S.; the development of trade agreements and the growth of international trade; and the proliferation of different quality standards throughout the world, including both product specifications and quality management systems.

The world markets grew rapidly during the 1970s and the 1980s but were characterized by products and services that varied widely in their performance, characteristics, styling, materials, and interchangeability of parts. Since a key aspect of the ISO mission is to facilitate trade and remove trade barriers, ISO formed the Technical Committee (TC) 176 in 1979 to address these issues under the general topic of quality management. "The goal was to make it possible for purchasers in the international marketplace to ensure that products they bought were manufactured in accordance with known, verifiable, and accepted methods of controlling the manufacture and distribution of products" (Bell 1995).

TC 176 was confronted by a bewildering array of quality standards for both product characteristics and quality management programs that had been institutionalized by various industrialized nations throughout the world. Such diverse standards constituted a set of technical barriers to trade. A wide variety of interpretations of the same standards, both within an industry and between industries, caused problems with products manufactured and traded worldwide. Industries generally agreed, however, that product specifications were not enough. Quality management had to address the processes used for purchasing, producing, and servicing to ensure attainment of uniform quality levels.

TC 176 set about the task of harmonizing the various standards related to quality management systems throughout the world and issued its quality management and quality assurance standards, the ISO 9000 series, in 1987. Since its original publication, the ISO 9000 series has undergone three significant revisions in 2000, 2008, and 2015. This series addresses the processes used by a business to ensure that it meets customer requirements for its products and services. The 9000 series does not define specific product performance levels or physical characteristics but describes how to manage a business with a quality focus, thus attaining consistent results and providing confidence to the customer.

The 9000 series, as the forerunner of the 14000 series, has proved to be extremely popular throughout the world. Currently, more than 1 million businesses in more than 170 countries worldwide have achieved ISO 9001 certification. Worldwide, achievement of ISO 9000 registration has become a prerequisite for doing business in dozens of countries. Some governments have even made the ISO 9000 series mandatory for businesses in their countries. Those businesses wishing to enter the European markets need to consider an investment in an ISO 9000 quality management system as an essential component of their business. ISO 9000 is a legal requirement in the medical devices market in the European Union (EU). It is also required in many other markets where product risk is a factor, for example, high-pressure valves and public transportation. Competitive pressure is the primary reason for adopting ISO 9000, as thousands of firms are placing themselves on preferred supplier lists and demonstrating their global commitment to quality.

As not only quality management but also risk management increased in significance, in 2011 ISO responded to market needs by establishing TC 262 to address standardization of risk management practices. This wide-ranging body currently includes 55 participating countries and 18 observing countries and is responsible for the following family of standards:

- ISO 31000—Risk management—Guidelines
- ISO 31010—Risk management—Risk assessment techniques
- ISO 31022—Risk management—Guidelines for the management of legal risk
- ISO 31030—Managing travel risks—Guidance for organizations
- ISO 31050—Guidance for managing risks to enhance resilience
- ISO Guide 73—Risk management—Vocabulary

The goal of these standards, as stated on the ISO/TC 262 web page, is "to support organizations in all their activities including making decisions to manage and minimize the effects of accidents, disasters and faults in technical systems as well as response and recovery from major disruptive risks." TC 262 also continues to explore new areas in which to provide support.

1.2 ISO 14000 Series of Standards for Environmental Management Systems

Many of the trends that resulted in the emergence of the 9000 series also played a part in the development of the 14000 series. These trends include the growth of international markets, the proliferation of environmental management standards and regulations in various countries, and the adoption of environmental management programs by businesses in response to complex environmental regulations.

Several issues related to the formation of the ISO 14000 series were different from those of ISO 9000. Any attempt to standardize environmental performance worldwide involves considerable social and political controversy. Implementing environmental management systems in countries with social democracies, such as Norway, has been relatively easy. In countries such as the U.S., the often antagonistic and litigious relationship between the government and the regulated community has in the past caused environmental issues to be approached by business people with extreme caution, if not fear (Begley 1996).

Two trends in environmental management have emerged as the driving forces for ISO 14000. In 1972 the United Nations held a conference on the environment in Stockholm, Sweden. Later environmental conferences were held in Rio de Janeiro in 1992 and 2012, and in Paris in 2015, as well as summits in Johannesburg in 2002 and New York in 1997 and 2015. The global community came together at each gathering to meld the views of diverse and sometimes opposing groups into a firm commitment to responsible environmental management and global sustainability. At the end of 2015, the Paris Agreement was negotiated by 196 parties of the United Nations Framework Convention on Climate Change (UNFCCC) and was adopted by consensus. For the first time, the world established the environment as a priority in national and international affairs.

A seemingly opposing force, exemplified by U.S. President Donald Trump's 2019 withdrawal of the U.S. from the Paris Agreement, had already coalesced by 1986 in the Uruguay Round of the General Agreement on Trade and Tariffs (GATT). These negotiations resulted in a commitment to foster international trade. The Agreement on Technical Barriers to Trade section of the GATT does, however, encourage the use of international standards and conformity to assessment systems in order to improve the efficiency of production and facilitate trade. This treaty requires that these international standards not be prepared, adopted, or applied with a view to, or with the effect of, creating unnecessary obstacles to international trade. Indeed, the technical regulations cannot be more trade restrictive than necessary to fulfill a legitimate objective such as national security requirements; the prevention of deceptive practices; or the protection of human health or safety, animal or plant life or health, or the environment. This treaty establishes what could be perceived as an anti-environmental force by positioning international trade as a competing priority with environmental protection.

As a response to these emerging trends in environmental management, and cognizant of the resounding success and worldwide adoption of the ISO 9000 standards on quality management systems, in August 1991 ISO established a Strategic Advisory Group for the Environment (SAGE). This advisory group's purpose was to assess the need for environmental management standards and to recommend an overall strategic plan to develop these standards. ISO requested SAGE to consider the following issues:

- Promote a common approach to environmental management similar to total quality management standards (ISO 9000)
- Enhance businesses' abilities to attain and measure improvements in environmental performance
- Utilize international standards to facilitate trade and remove trade barriers

SAGE was specifically instructed not to consider environmental criteria, such as levels of pollutants, health assessments/risks, technology specifications, or process criteria. For over a year, SAGE studied the U.K.'s BS 7750 and other national environmental management standards as possible starting points for an ISO version. In 1993 SAGE recommended the formation of an ISO technical committee dedicated to the development of a uniform international environmental management standard, as well as other standards on environmental management tools. ISO formed Technical Committee (TC) 207 to develop a series of standards on environmental management systems to accomplish standardization in the field of environmental management tools and systems.

In June 1993, TC 207 met for the first time in Toronto, Canada, where some 200 delegates representing approximately 30 countries agreed to complete a draft of the environmental management standard and international auditing standards. SAGE was officially disbanded. Following an interim meeting from April 17 to 20, 1994, in Surfer's Paradise, Australia, TC 207 met again from June 24 to July 1, 1995, in Oslo, Norway, where 600 delegates representing over 50 countries agreed to elevate the environmental management standard and auditing standards to draft international standards with a scheduled publication by the end of 1996. By July 1995, TC 207 had members from 63 countries.

Six technical subcommittees (SCs) and working groups (WGs) were created within TC 207 and are currently formulating standards in the following areas:

- **SC1: Environmental Management Systems**, with the U.K. as the secretariat, administered by the British Standards Institution
- **SC2: Environmental Auditing and Related Environmental Investigations**, with the Netherlands as the secretariat, administered by the Netherlands Normalisatie-Instituut
- **SC3: Environmental Labeling**, with Australia as the secretariat, administered by Standards Australia
- **SC4: Environmental Performance Evaluation**, with the U.S. as secretariat, administered by ANSI
- **SC5: Life-Cycle Assessment**, with France as the secretariat, administered by Association Francaise de Normalisation
- **SC6: Terms and Definitions**, with Norway as the secretariat, administered by Norges Standardiseringsforbund
- **WG1: Environmental Aspects in Product Standards**, with Germany as the secretariat, administered by the Deutsches Institut für Normung e.V.

From June 17 to 21, 1996, TC 207 delegates from 50 countries and 10 liaison organizations met in Rio de Janeiro, Brazil, to complete the ISO 14000 drafting process. The main outcomes of this meeting were the following:

- The 14001 and 14004 environmental management systems standards received final approval and were issued by ISO in September 1996. The 14001 environmental management systems specification was also approved.
- The 14010, 14011, and 14012 environmental auditing standards had been in the final stages of publication and were overwhelmingly approved in the plenary meeting. These standards were published in conjunction with the 14001 and 14004 standards in September 1996.

1.3 Principles Used in Developing International Standards

To understand how the 14000 series was developed and how the standards within this series have been applied, we need to look at the three basic principles that ISO follows in developing all standards: consensus, encouragement of full participation, and voluntary adoption.

Consensus is a difficult task when one considers the complexity and controversy surrounding environmental issues worldwide. The development process for ISO standards, however, ensures that consensus is achieved through these basic steps:

Preparing a Justification. An appointed committee, usually part of a subcommittee, prepares a justification for any proposed standard and formally submits a New Work Item proposal to the entire technical committee or the relevant subcommittee for a vote. If the majority of participating members vote in favor of the proposal and at least five members declare their commitment to actively support the project, development of the new standard proceeds.

Preparation of Working Documents. Groups of experts develop working drafts of the new standard and refine them so they can advance to the next phase. It is in this phase that controversies are discussed, contentious issues are resolved at the most basic level, and a general consensus is reached among the experts.

Preparation of Committee Drafts. The working documents are formalized as Committee Drafts (CDs) and distributed to the entire technical committee for comments. Comments are reviewed and included, and as many CDs as necessary produced until a consensus is reached among the technical committee members and the document is ready to proceed to Draft International Standard (DIS) status.

Preparation of Draft International Standard. The DIS is circulated to all ISO member bodies for voting and comments within a period of six months. If a two-thirds majority of the participating members approves the standard and not more than one quarter of the members disapproves it, then the standard is moved forward to publication.

Consensus is achieved first among the technical experts, then within the technical committee, and finally among all the standards organizations and their experts throughout the world. Thus, consensus is built in as a key aspect of the standards development process.

Participation in technical committees is open to all qualified and interested individuals. Industries that are likely to be affected by the standards are often involved with ISO technical committees as members of their national standards bodies. Adoption of these standards by countries and industries is voluntary and based on market-driven needs. Thus, any businesses anywhere in the world can voluntarily choose to adopt these standards if it helps them to attain their vision or pursue their mission, based on the judgment of management. Businesses and their managers can be assured that the standards contain a high degree of technical integrity, resulting from the consensus of both industry experts and national standards bodies.

1.4 Goals of ISO 14000

The goal of ISO 14000 is to evolve a series of generic standards that provide business management with a structured mechanism to measure and manage environmental risks and impacts. Major components of a strong environmental management system that are reflected in the ISO 14001 standard and others include

> (1) top management leadership, policy setting, and review; (2) identifying environmental issues/risks (or "aspects" in the vocabulary of ISO 14001) and legal and other requirements; (3) establishing objectives to successfully manage these risks and requirements in line with policy; (4) implementing programs and procedures (including for emergencies) that establish how, by whom, and when those objectives will be met; (5) training people so that they know their risks and what to do to best manage them; (6) monitoring, measuring, and auditing to track performance and verify implementation; (7) taking preventative and corrective action; (8) continual improvement; and (9) documentation and record keeping.
>
> **(Bell 2018)**

Standards have been published or are being developed for:

- Environmental management systems (ISO 14001–14002; 14004–14009)
- Environmental auditing and related environmental investigations (ISO 14015–14016)
- Environmental labeling (ISO 14020–14027)
- Environmental performance evaluation (ISO 14030-01, -02, -03, -04; 14031; 14033; 14034; 14063)
- Life-cycle assessment (ISO 14040; 14044–14049; 14071–14073)
- Greenhouse gas management and related activities (ISO 14064–14067; 14069; 14080; 14082; 14090–14092; 14097)

These standards are designed to help a business establish and meet its own policy goals through objectives and targets, organizational structures and accountability, management controls and review functions—all with top management oversight. The focus is on management rather than on performance standards. The centerpiece of this section, the 14001 standard for environmental management systems, provides a framework for assessing, managing, and reducing the liabilities associated with environmental aspects of operations. Through several key requirements of the 14001 standard, environmental management becomes a strategic decision-making concern, allowing management to make more effective decisions for reducing risks by reducing their environmental impacts and risks.

ISO 14000 is only one of the many efforts being undertaken by businesses, environmental groups, investor organizations, governments, and other entities to improve the environmental performance of businesses. Nonetheless, ISO 14000 has a unique role because, as an international standard, it will be given more credence than its more narrowly focused counterparts and garner universal acceptance. For this reason, among others, it already fulfills an important function in the development of environmental management systems around the world. As subsequent chapters demonstrate, other entities also have had significant impacts on the development of environmental risk management and reduction systems, and they must be carefully considered in conjunction with ISO 14000.

While ISO was considering the development of a series of international standards for environmental management, pressures were mounting from worldwide environmental organizations, like Greenpeace, and national organizations, like the Sierra Club, the Natural Resources Defense Council, the Environmental Defense Fund, and others, to make businesses more responsible for their environmental impacts. Professional organizations began developing their own guidelines on good environmental management practices. The Chemical Manufacturers Association created its Responsible Care Program, which was adopted by all its member companies to improve the safety of handling and disposal of chemicals and otherwise to reduce environmental impacts. The Global Environmental Management Initiative (GEMI) developed and published its own set of guidelines based on total quality environmental management principles and addressed such topics as cost-effective pollution prevention, environmental reporting, environmental health and safety training, and benchmarking. The International Chamber of Commerce (ICC) published its own set of environmental principles for adoption by its members and other organizations to improve environmental decision making and reduce impacts. Responding proactively to what they recognized as "global megatrends"—namely, increasing competition for finite resources and fast population growth—that are "challenging traditional notions of growth and development," the ICC initiated their International Charter for Sustainable Development in 1991 and released updated versions in 2000 and 2015. The ICC's stated aim in providing the 2015 Charter, which draws from their Green Economy Roadmap, is

To provide a practical framework including tools for businesses of all sectors and geographies to help them shape their own business sustainability strategy. In doing so, it is also intended to be relevant for small and medium-sized companies and businesses in emerging markets as a common and accessible starting point.

It details a framework for businesses to establish management systems that consider environmental, economic, and societal aspects of business. This triple bottom line strategy is vital in guiding and integrating sustainability into daily business practices. Even local business organizations began to get proactive. For example, the Denver Metro Chamber of Commerce developed its own environmental policy statement and a tool kit for its members to create their own environmental policies.

Other organizations decided to apply social, political, and investment pressures to change business practices. This chapter highlights one public interest group, the Coalition for Environmentally Responsible Economies (CERES), that decided to force businesses to change their corporate cultures. The preliminary results of its campaign are compared with those of the National Center for Preventive Law, an organization that assembled a group of experts to create voluntary corporate compliance principles and guidelines for businesses.

1.5 CERES

In 1989, an organization known as the Social Investment Forum brought together various environmental groups, institutional investors, government agencies, and economists to make businesses adopt principles to govern their environmental performance. This group formed the Coalition for Environmentally Responsible Economies. The organizers believed that a collaboration between institutional investors and environmental groups could have a significant positive impact on the environment by pressuring businesses to reduce their environmental impacts. CERES has served as an impetus for getting businesses to adopt their own voluntary environmental management programs.

On September 7, 1989, CERES issued ten principles, originally known as the Valdez Principles. These principles were named after the Exxon Valdez tanker that ran aground on March 24, 1989, spilling 11 million gallons of crude oil into the waters of Alaska's Prince William Sound. CERES modeled its Valdez Principles after the Sullivan Principles, written in 1977 by Reverend Leon H. Sullivan of Philadelphia in an effort to promote social justice and eliminate apartheid in South Africa. Within five years, more than 140 multinational corporations, including Exxon, IBM, Citicorp, and Mobil, signed the Sullivan Principles. In contrast, the CERES Principles, which evolved from the Valdez Principles, have been adopted by about 60 companies, including only 13 Fortune 500 companies, as of early 2019.

The Sullivan Principles required companies doing business in South Africa to desegregate work facilities, provide equal employment practices for all employees, initiate training programs for minorities, promote minorities to management and supervisory positions, and improve the quality of employees' lives outside the work environment. In 1985 U.S. President Ronald Reagan signed an executive order that included a requirement that American businesses in South Africa abide by the Sullivan Principles. No president has issued an executive order endorsing the CERES Principles; however, regional offices of the Environmental Protection Agency (EPA) have been supportive of CERES' effort.

CERES proponents believed that social investment forces could be marshaled to cause businesses to sign the Valdez Principles in much the same way as the multinational businesses signed the Sullivan Principles. CERES drafted its principles to address the release of pollutants, the sustainable use of natural resources, reduction and disposal of waste, energy efficiency, and conservation and risk reduction to employees and surrounding communities. CERES' objective was "to create a voluntary mechanism of corporate self-governance that will maintain business practices consistent with the goals of sustaining our fragile environment in future generations, within culture that respects all life" (Bavaria 1989). CERES proponents believed that businesses who signed on would be recognized as leaders in making a voluntary public commitment to environmental accountability.

According to the CERES proponents, there are four advantages to businesses that agree to adopt the Valdez Principles. Companies can (1) receive positive publicity that improves their corporate image in an age of "green consumerism"; (2) reduce waste-hauling fees and increase revenues generated by recycling; (3) strengthen their environmental standards and avoid financially devastating environmental disasters; and (4) receive investments from CERES members. CERES asked businesses to make a "quasi-legal commitment" by signing the Principles. Each year thereafter, signatory businesses would be required to submit detailed performance information setting forth how well they were complying with the mandate of the Principles. CERES planned to hire auditors annually to independently verify the businesses' self-audits and then grade each business against each Principle to provide an overall CERES score. CERES intended to disseminate the information as widely as possible, perhaps by publishing the environmental equivalent of Consumer Reports. The original Valdez Principles stated:

VALDEZ PRINCIPLES

PRINCIPLE 1: PROTECTION OF THE BIOSPHERE

We will minimize and strive to eliminate the release of any pollutant that may cause environmental damage to the air, water or earth or its inhabitants. We will safeguard habitats in rivers, lakes, wetlands, coastal zones and oceans and will minimize contributing to the greenhouse effect, depletion of the ozone layer, acid rain, or smog.

PRINCIPLE 2: SUSTAINABLE USE OF NATURAL RESOURCES

We will make sustainable use of renewable resources, such as water, soils, and forests. We will conserve non-renewable natural resources through efficient use and careful planning. We will protect wildlife habitat, open spaces, and wilderness while preserving biodiversity.

PRINCIPLE 3: REDUCTION AND DISPOSAL OF WASTE

We will minimize the creation of waste, especially hazardous waste, and, wherever possible, recycle materials. We will dispose of all wastes through safe and responsible methods.

PRINCIPLE 4: WISE USE OF ENERGY

We will make every effort to use environmentally safe and sustainable energy sources to meet our needs. We will invest in improved energy efficiency and conservation in our operations. We will maximize the energy efficiency of products we produce or sell.

PRINCIPLE 5: RISK REDUCTION

We will minimize the environmental, health and safety risks to our employees and the communities in which we operate by employing safe technologies and operating procedures and by being constantly prepared for emergencies.

PRINCIPLE 6: MARKETING OF SAFE
PRODUCTS AND SERVICES

We will sell products or services that minimize adverse environmental impacts and that are safe as consumers commonly use them. We will inform consumers of the environmental impacts of our products and services.

PRINCIPLE 7: DAMAGE COMPENSATION

We will take responsibility for any harm we cause to the environment by making every effort to fully restore the environment and to compensate those persons who are adversely affected.

PRINCIPLE 8: DISCLOSURE

We will disclose to our employees and to the public incidents relating to our operations that cause environmental harm or pose safety or health hazards. We will disclose potential environmental, health or safety hazards posed by our operations, and we will not take any action against employees who report any condition that creates a danger to the environment or poses health and safety hazards.

PRINCIPLE 9: ENVIRONMENTAL DIRECTORS AND MANAGERS

At least one member of the Board of Directors will be a person qualified to represent environmental interests. We will commit management resources to implement these Principles, including the funding of an office of vice president for environmental affairs or an equivalent executive position, reporting directly to the CEO, to monitor and report upon our implementation efforts.

PRINCIPLE 10: ASSESSMENT AND ANNUAL AUDIT

We will conduct and make public an annual self-evaluation of our progress in implementing these Principles and in complying with all applicable laws and regulations throughout our worldwide operations. We will work toward the timely creation of independent environmental audit procedures which we will complete annually and make available to the public.

1.5.1 The Response from Business in the U.S.

Unlike the Sullivan Principles, which gained fairly swift approval by multinational corporations, no major business in the U.S. signed the Valdez Principles when they were first announced. From September 1989 until April 1990, CERES solicited comments from more than 150 businesses on their views on the Valdez Principles. CERES also sent out approximately 3,000 invitations to sign the Valdez Principles. This mailing list included the Fortune 1000 companies.

A common response was that businesses were disappointed that CERES had not requested their input during the initial drafting process. Some objected to signing the Principles because they did not wish to subject their individual environmental and business practices to review and oversight by an independent special interest group. Nor did they wish their environmental performance to be audited and graded by environmental activists. One of CERES' important and noteworthy goals was to actually build an environmental consensus among a broad spectrum of interest groups and to lessen the adversarial relationship between industry and environmental groups. This goal was met with skepticism from many of the companies that were invited to sign the Principles.

Nevertheless, some businesses were interested in reviewing the Principles and using them, as well as other information, to draft their own environmental policy statements without some of the more onerous public reporting requirements of CERES. Chevron dispatched representatives to talk with CERES and then, like many other major businesses, decided not to sign the Principles.

In April 1990, CERES issued a 22-page document titled *The 1990 Ceres Guide to the Valdez Principles*. This guide was intended to address corporate concerns and identify particular actions a business would be required to perform if it chose to sign the Principles. Neither the guide nor persistent efforts by CERES members, however, resulted in a groundswell of corporate support for the Valdez Principles. CERES membership grew slowly to only 42 businesses—mostly small businesses and CERES members that had originally signed the Principles when they were first issued. Businesses that signed on included Vermont ice cream makers Ben & Jerry's, Domino's Pizza Distribution, a lumber business, a paper recycling company, and two natural-ingredient personal-care products businesses. Collectively, these business operations did not significantly impact the environment.

CERES proponents and other environmental activists who were dissatisfied with the corporate response increased social and political pressure on businesses to sign on. Shareholders' resolutions were filed with more than 50 corporations, urging them to adopt the Valdez Principles. Only five were voted on in 1990. That number increased to 31 in 1991. On May 24, 1991, at General Motors' annual meeting in Nashville, Tennessee, Chairman Robert C. Stempel argued against the business adopting such a resolution:

> For more than three decades, General Motors has seen a clean and healthy environment as a top priority. We take pride in our own leadership role in reducing emissions from both vehicles and plants and in our work to minimize wastes and to dispose of those wastes in an environmentally sound manner.

On the strength of these arguments, the GM resolution was defeated. By September 1992, similar resolutions failed at 56 other stockholders' meetings.

The shareholder campaign began to dwindle, and from a record 54 resolutions filed in 1993, only 16 were submitted in 1996.

Major businesses found several significant reasons not to adopt the Valdez Principles. Initially, businesses were troubled by the inclusion of a combination of ambiguous language and broad terms that might subject corporations to a bottomless snake pit of liabilities. They feared that legal actions could easily be taken against a business that has an unsuccessful environmental policy which states that it goes beyond existing laws. Some claimed that adopting the Principles might cause substantial and unjustified increased operating costs. Others felt they mandated disclosures that might interfere with the confidentiality of new product development.

Amoco expressed the views of many corporations that opposed being forced to appoint an "environmental director." It released a statement that "[s]pecial interest directorships, environmental or otherwise, are ... bad policy from the standpoint of corporate governance." Some state legislatures, however, have actually endorsed the use of constituency directors by enacting statutes to allow corporate Boards to be concerned with other interests, including the environment. Tom Smith, Vice President of Public Affairs for Dow Chemicals, also opposed environmental directorships and questioned: "Who is qualified to — quote 'represent environmental interests'?" John McCallister of DuPont added that board members must "represent stockholders, not constituencies."

Some businesses refused to sign the Valdez Principles because they believed that releasing extensive environmental reports to CERES might lead to undue outside influence and pressure on how they conducted internal environmental affairs. One activist predicted as much by advising CERES to establish reasonable benchmarks first so that organizers could recruit more businesses. He urged CERES to get businesses to sign on to the Principles and gradually increase the stringency of the standards. He argued that businesses would be reluctant to abandon the code once they have publicly signed.

CERES did not follow these bait-and-switch tactics. Instead, CERES thoughtfully listened to corporate complaints, modified some of the Principles, and changed the name from the Valdez to the CERES Principles. CERES eliminated the environmental director requirement and replaced it with the statement, "[i]n selecting our Board of Directors, we will consider demonstrated environmental commitment as a factor." Also, a disclaimer was added to alleviate corporate concerns that the Principles would be used against endorsers in litigation:

> [t]hese principles are not intended to create new legal liabilities, expand existing rights or obligations, waive legal defenses, or otherwise affect the legal position of any endorsing company, and are not intended to be used against an endorser in any legal proceeding for any purpose.

Other revisions were made to make the Principles more palatable to businesses. For example, *Principle 5* ("We will minimize the environmental, health and safety risks to our employees and the community ... by being constantly prepared") was revised to read, "We will strive to minimize ... by being ... prepared." *Principle 7* ("We will take responsibility for any harm we cause to the environment by making every effort to fully restore the environment and to compensate those persons who are adversely affected") was changed to "We will promptly and responsibly correct conditions we have caused that endanger health, safety or the environment. To the extent feasible, we will address injuries we have caused to persons or damage we have caused to the environment and will restore the environment" (CERES Principles).

These significant revisions resulted in progress for CERES. On February 10, 1993, Sun Company, Inc. became the first Fortune 500 company to endorse the CERES Principles. Shortly thereafter, H.B. Fuller, a specialty chemical manufacturer, signed on—followed over the course of three years by General Motors, Polaroid, Bethlehem Steel, and BankAmerica. After eight years' effort, however, only 50 businesses, including one Boston law firm, had endorsed the CERES Principles. Many businesses that decided not to endorse the Principles nonetheless saw the wisdom of implementing effective environmental management systems. These businesses have found it wiser and more productive to design policy statements that fit their own individual needs and advance their goals and objectives in the marketplace and the worldwide environment. Instead of adopting the mandatory self-audit and self-disclosure procedures designed by CERES, businesses have found their own way to manage their environmental concerns and present their environmental records to the public.

For their part, CERES, now rebranded as Ceres, adapted by changing their focus from promoting their Principles to networking (with the formation of Investor, Company, Policy, and Nonprofit Networks) and establishing initiatives such as Commit to Climate, Climate Action 100+, We Are Still In, Connect the Drops, and Clean Trillion. Ceres has thus continued to be a positive influence in motivating businesses to adopt policies and evaluate and improve their environmental performance.

1.5.2 Ceres' Impacts

It is hard to quantify Ceres' overall impacts on corporate environmental performance, including self-reporting of environmental results. One of its proactive endorsers, Polaroid, produced one of the first environmental reports four years before it signed the CERES Principles. Recent statistics show that there is a growing trend to report. Approximately 120 large U.S. companies, or 20% of selected industries from the S&P 500 and Fortune 500, issued a corporate environmental report by 1995 (Lober 1997). By 2011, that figure still held at roughly 20% of S&P 500 companies. By 2012, however, 53% of S&P 500 companies issued a sustainability or corporate responsibility report, and by 2016, the figure had risen to 82% of S&P 500 companies (Governance & Accountability Institute 2017). In a sample size of 4,900 large and mid-cap companies from around the globe, Swiss professional services company KPMG found that 75% of the companies issued a corporate responsibility report in 2017, with a greater than 60% reporting rate in every single business sector represented in the study. They reasonably concluded that corporate responsibility reporting has become "standard practice" for large and mid-cap firms. Significant reporting increases were particularly seen in Latin America and in countries with relevant new regulations, such as Mexico, New Zealand, and Taiwan (KPMG 2017).

While environmental reports are highly varied in format, the introduction of reporting guidelines in 1997 and new standards in 2016 by the Global Reporting Initiative (GRI) has begun helping to create a more uniform reporting framework. KPMG found that, by 2017, 63% of the 4,900 businesses surveyed utilized GRI guidelines or standards, with most of those still utilizing the GRI G4 guidelines (KPMG 2017). GRI reports that 93% of the world's largest 250 corporations utilize GRI standards to report on their sustainability performance (GRI). In addition, third-party assurance of reported data has been steadily increasing, with about half of the reporting companies globally having sought it out in 2015–2016. North American companies lag behind on this front, however, with fewer than one third of the 551 American and Canadian companies studied by the Centre for Sustainability and Excellence (CSE) in 2015–2016 having sought out third-party assurance of reported data (CSE, 2017). And while both KPMG and CSE found that only about 40% of firms linked their environmental responsibility activities to the UN's Sustainable Development Goals (adopted in 2015 with the stated aim of being achieved by 2030), they assessed such explicit linkage to be a steadily growing trend in environmental reporting.

Undoubtedly, many businesses are preparing their own annual corporate environmental reports because they are proud of their environmental accomplishments. The 2017 CSE study points to the rewards of environmental reporting, as they found that "about two thirds of companies with the highest rankings on sustainability ratings such as CSRHub had better financial performance than companies with lower rankings as indicated by revenue during the period 2014–2016 than those without reports" (CSE founder Nikos Avlonas, quoted in Waghorn 2017).

With environmental reporting becoming increasingly de rigueur (see subsequent chapters for specific examples of environmental reports), it is remarkable how popular this method of benchmarking environmental performance has become in so short a period of time. Ceres' request to companies to make a public disclosure of their environmental performance has had some effect on businesses deciding to become more proactive in disclosing their performance to the public. Other organizations have taken a more direct and confrontational approach to changing corporate environmental behavior.

1.6 NCPL

In 1994 Edward A. Dauer, president of the National Center for Preventive Law (NCPL), brought together 33 experts from a broad variety of fields, including corporate risk management, law, education, and communications. Unlike Ceres, these proponents of corporate compliance were not just interested in environmental protection. Motivated in part by the U.S. Federal Sentencing Guidelines' directive that companies can reduce their exposures by adopting a compliance program, this group viewed their mission broadly—to assist companies to prevent and detect violations of any type of corporate law. The main areas of concern were antitrust and other fair-trade laws, government procurement and contracting, political contributions and lobbying, securities and insider trading, money laundering, environmental issues, labor relations and employment discrimination, sexual harassment, intellectual property, substance abuse, product liability, consumer protection, workplace safety, conflicts of interest, and commercial bribery. Over the course of two years, this distinguished group of experts met and discussed how to achieve superior corporate compliance performance by creating effective compliance programs. The group's goal was to draft a unified and comprehensive set of principles and general guidelines for corporations to help them achieve compliance with civil and criminal laws. The Commission's Principles are as follows:

NATIONAL CENTER FOR PREVENTIVE LAW
CORPORATE COMPLIANCE PRINCIPLES

ESTABLISHING COMPLIANCE PROGRAMS

Principle 1: Manage Compliance

Organizations should pursue compliance through the creation and maintenance of an effective compliance program.

Principle 2: Contain Risks

An effective compliance program is designed to prevent, detect, and respond to legal risks and to promote compliance with the law.

Principle 3: Respond to Change

An effective compliance program is a dynamic process that is designed to be flexible and modified, when appropriate, to reflect changing conditions.

Principle 4: State Compliance Policy

An effective compliance program states that it is the organization's policy to comply with all applicable laws.

Principle 5: Endorse at Top Levels

The highest governing authority within an organization should endorse the organization's compliance program.

Principle 6: Create Compliance Accountability

An effective compliance program establishes accountability for compliance throughout the organization.

Principle 7: Ensure Program Fairness

An effective compliance program is designed to operate fairly and equitably.

STRUCTURE AND CONTROL

Principle 8: Maintain High-Level Oversight

Specific high-level personnel in an organization are responsible for the administration and oversight of the compliance program.

Principle 9: Assign Individual Responsibility

A compliance program has the support of senior management of the organization. Each officer, manager, and employee is responsible for supporting and complying with the compliance program's standards and procedures.

Principle 10: Delegate Authority Responsibly

The organization exercises due diligence to prevent the delegation of substantial discretionary authority to persons having a propensity to engage in illegal activities.

Principle 11: Enforce Internally

The organization takes reasonable steps to achieve compliance with its standards and the law.

Principle 12: Reward Success

Incentives and disincentives are significant tools in promoting compliance.

COMMUNICATIONS AND TRAINING

Principle 13: Communicate Standards

The organization's compliance program has a communications component, the objectives of which are to make employees and other agents aware of applicable standards of conduct and to promote compliance.

Principle 14: Match Training to Tasks

An effective compliance program communicates appropriate compliance information and motivation to the organization's employees and other agents.

Principle 15: Tailor Training to Audience

An effective communications program is designed to reach the intended audience.

Principle 16: Define Communication Responsibilities

All levels of management are responsible for the operation of an organization's compliance communications program.

RESPONSES TO VIOLATIONS

Principle 17: Respond Proactively

An effective compliance program is proactive in its approach to dealing with incidents of noncompliance.

Principle 18: Gather Compliance Information

An effective compliance program possesses or has access to investigatory, evaluative, and reporting resources and utilizes those resources to monitor compliance.

Principle 19: Consider Offense Reporting

An effective compliance program addresses the occasions for external reporting of violations of the law.

Principle 20: Evaluate Program Effectiveness

An effective compliance program utilizes incidents of noncompliance to evaluate its own effectiveness, to correct deficiencies, and to effect improvements.

Source: Reprinted with permission from NCPL.

1.7 Two Approaches, One Goal

Ceres and NCPL took entirely different approaches to achieve the goal of improving corporate performance. Putting aside the advantages and disadvantages of each approach, all companies would agree that preserving corporate reputation is a significant reason to be proactive in this field.

A good corporate image, however, is only one reason to monitor and improve environmental performance. The satisfaction of saving money and minimizing the use of energy and natural resources by eliminating or diminishing environmental impacts is a powerful incentive for improving environmental performance. In the following chapters are a number of strategies for managing and reducing environmental risk. We focus initially on ISO 14000 because it presents a comprehensive systems approach to achieving better environmental results. Later chapters will discuss how such programs can be augmented by risk management and reduction strategies and initiatives. Far less confrontational than Ceres, ISO 14000 adopts the same approach as the NCPL Commission, allowing businesses to select the pace and timing of improvements to environmental performance in a global marketplace that is increasingly becoming more interconnected and concerned with environmental degradation. Properly implemented, ISO 14000 can result in a wide range of benefits and opportunities for significant environmental and economic gains.

ISO 14000 presents another alternative for companies to deal with their rigorous environmental responsibilities. The next chapter discusses how the practice of environmental risk management has changed and how ISO 14000 can play a key role in the development of systems that will ensure compliance with environmental laws and create opportunities to eliminate pollution and industrial waste in the future.

References

Bavaria, J. "Clean Up Your Environmental Act: Withholding Investments Can Influence Corporate Actions." *Newsday*, September 07, 1989.

Begley, R. "Is ISO 14000 Worth It?" *Journal of Business Strategy* 17 (September/October 1996).

Bell, C.L. "ISO 14001: Application of International Environmental Management Systems Standards in the United States." *Environmental Law Reporter*, December 1995.

Bell, C.L. "ISO 14001 and Environmental Management Systems: Where Are We?" *A&WMA, The Magazine for Environmental Managers*, July 2018.

Centre for Sustainability and Excellence. "Sustainability Reporting Trends in North America." 2017. https://www.cse-net.org/wp-content/uploads/documents/Sustainability-Reporting-Trends-in-North%20America%20_RS.pdf.

Governance & Accountability Institute. "Flash Report: 82% of the S&P 500 Companies Published Corporate Sustainability Reports in 2016 [Press release]." 2017. https://3blmedia.com/News/Flash-Report-82-SP-500-Companies-Published-Corporate-Sustainability-Reports-2016.

GRI. "About GRI." https://www.globalreporting.org/Information/about-gri/Pages/default.aspx.

Hall, Jr., R.M. "ISO 14000 Environmental Management Standards: Making the Benefits Outweigh the Burdens." *Paper Presented at the 25th ABA Annual Conference on Environmental Law*, Keystone, CO, March 21–24, 1996 (on file with the authors).

International Chamber of Commerce. "ICC Business Charter for Sustainable Development 2015." n.d. https://iccwbo.org/publication/icc-business-charter-for-sustainable-development-2015/.

ISO. "ISO/TC 262." https://committee.iso.org/home/tc262.

KPMG. "The Road Ahead: The KPMG Survey of Corporate Responsibility Reporting 2017." 2017. https://assets.kpmg/content/dam/kpmg/xx/pdf/2017/10/kpmg-survey-of-corporate-responsibility-reporting-2017.pdf.

Lober, D.J. "Current Trends in Corporate Reporting." *Corporate Environmental Strategy*, Winter 1997.

Murphy, C.N., and J. Yates. *The International Organization for Standardization (ISO): Global governance through voluntary consensus*. New York: Routledge, 2009.

National Center for Preventive Law. *National Center for Preventive Law Corporate Compliance Principles*, 1996. (For a complete copy of this document, please contact: NCPL, 1900 Olive Street, Denver, CO 80220, Ph. 303-871-6099.)

United Nations. "Sustainable Development Goals." https://www.un.org/sustainable development/sustainable-development-goals/.

Waghorn, T. "Sustainable Reporting: Lessons from the Fortune 500." *Forbes*, 2017. https://www.forbes.com/sites/terrywaghorn/2017/12/04/sustainable-reporting-lessons-from-the-fortune-500/#4718c6ca6564.

2

A Systematic Approach to Managing Environmental Risk

Robert A. Woellner

Environmental risk management finally came of age at the end of the 20th century. This field matured in the 1980s and 1990s through the incorporation of the diverse elements of communications, toxicology, insurance, litigation avoidance, and regulatory compliance. In earlier years, there had never been a coordinated, holistic approach to environmental risk management—one that integrated these seemingly disparate disciplines into a meaningful structure to help senior management make risk-based decisions. The ISO 14000 series of standards for environmental management presented a uniform approach upon its publication in 1996. ISO 14000, including the updated ISO 14001 that was published in 2015, attempts to solve the problems that have made environmental risk management more technical than strategic, more reactive than proactive.

2.1 Responses of Companies to the Regulatory Arena

The U.S. Congress has passed a number of environmental laws designed to clean up pollutants and waste disposed of in the past and to effectively control their future disposal. Congress has delegated the responsibility to promulgate regulations that achieve the broad legislative objectives of the Clean Water Act, the Clean Air Act, the Safe Drinking Water Act, the Resource Conservation and Recovery Act (RCRA), the Comprehensive Environmental Response, Compensation and Liability Act (CERCLA), and many others to the U.S. Environmental Protection Agency (EPA). The EPA, in turn, has issued thousands of pages of complex regulations to protect the environment. No one is suggesting that all of these laws and regulations are flawed, or that they have not had beneficial effects for the country. Indeed, the EPA estimated in 2011 that the documented and projected health benefits of the Clean Air Act from 1990 to 2020 are approximately $2 trillion at a cost of $65 billion to industry (EPA, 2011). It is the price we, as the public, pay for the regulations, while their effectiveness is still in doubt. Many people believe that businesses have been too focused on responding to regulatory agencies and lawsuits, rather than on solving their own environmental problems.

Four out of five environmental rules created by the EPA are challenged in court. (It should be noted that presently, owing to the perceived anti-regulatory stance and lack of transparency of the EPA under the current administration, the majority of lawsuits filed against the EPA have been brought to combat secrecy or regulatory delays and rollbacks, rather than the establishment of regulations. It remains to be seen where this sea change will lead.) In the 1970s and 1980s, large companies created separate environmental divisions under environmental managers to respond to command-and-control directions from environmental authorities. These managers assumed the complicated and unenviable task of ensuring that their businesses complied with all the laws. Unfortunately, some believed that senior management should not be bothered with the details of environmental compliance. From the 1970s until recent times, many business owners and senior managers (with some notable exceptions such as recently retired Unilever CEO Paul Polman and others) deferred to their environmental managers to handle environmental matters concerning the business and did not involve themselves unless major problems arose. Rigorous environmental regulations resulted in end-of-pipe solutions that sought to catch pollution at the end of the process, rather than focusing on their effective management. Additionally, the overly litigious nature of the U.S., and increasingly other countries, has resulted in tremendous amounts of money being spent on defending lawsuits with no positive effect on improving the environment. It has become obvious to people involved in the command-and-control regulatory framework that the current process is ineffective and unmanageable. Conducting business and government as usual is a costly and divisive method of protecting the environment.

CERCLA is one of the best examples of a well-intentioned environmental legislation that has failed to achieve its goals. In 1980 Congress predicted that environmental remediation under CERCLA would ultimately cost several billion dollars. By 1989 the estimated cost for the remediation of hazardous waste sites had increased to $300 billion. In 2010, with funding for CERCLA having been reduced, the Government Accounting Office reported that "EPA's estimated costs to clean up existing contaminated sites exceed the Superfund program's current funding level" (Government Accounting Office 2011). By 2011, it was estimated that $335 million–$681 million per year would be needed for CERCLA in the coming years. Much of the money spent on these sites has gone to lawyers and consultants for transactional costs, including challenges to the scope and necessity of the cleanups. The government, businesses, and the environmental community are increasingly reaching a consensus that, to be most effective in their efforts to clean up and preserve the environment, businesses need to employ proactive management techniques to prevent contamination.

These efforts have been hampered by inadequate data generated by current internal environmental management practices that rarely present a systematic picture of organizational operations and, in most cases, are not adaptable to strategic decision making. Environmental data can be difficult to piece together to achieve an overall picture of where the business stands with regard to the environmental risks it poses.

Environmental profiles and risk-related information are often pushed to the lower levels of management, where the sheer volume of data is difficult to interpret and manage. These data are generally not utilized to make better management decisions, nor are they considered in the allocation of scarce organizational resources. The end result is that environmental risks are generally ignored until they approach extreme levels, at which point senior management becomes involved. For many businesses, however, this is too late, and their operational and financial performances suffer as they try to discover quick fixes for problems that have evolved over many years.

2.2 ISO 14000 Systems Approach

The ISO 14000 series of standards changes this scenario considerably. The prominence of formal environmental management systems grew, in large part, out of the ISO 14001 standard, which has become the most widely used EMS model throughout the world. Whether implemented for certification (of which there are more than 350,000) or used as a guide by many companies that do not seek third-party certification, ISO 14001–styled environmental management systems play a crucial role in devising sustainable development strategies. ISO 14000 standards are woven into the fabric of environmental compliance, with ISO 14001 having been recognized by the US EPA and many states as a framework for effective compliance programs (Bell 2018).

First, the standards mandate that environmental management be among the highest corporate strategic priorities. The standards provide a framework for top management to assess, manage, and reduce the risks associated with the environmental aspects of operations. Through several key requirements, the ISO 14000 standards elevate environmental management to areas of strategic decision making where, from an organizational perspective, more effective decisions for reducing risks can be made. Companies may no longer delegate environmental decision making only to low-level management personnel who operate in an unchecked or unverified environment.

The ISO 14001 standard for environmental management outlines an organizational framework for the systematic identification, control, and improvement of all environmental *aspects* and *impacts*. Rather than focusing exclusively on environmental performance, the standard addresses the management systems used to control environmental performance. This is a marked departure from the command-and-control approach commonly utilized by state and federal environmental protection agencies to regulate business activities. Management is still accountable for attaining emission levels prescribed by regulations and agreements. The integrated series of management systems in ISO 14000, however, motivate and encourage management to actively seek ways to reduce or eliminate pollutants and go beyond compliance.

The ISO 14001 standard for environmental management systems requires businesses to follow five basic and logical activities:

- Establish senior-management-level commitment to environmental management and promulgate a comprehensive environmental policy
- Develop objectives, targets, and a program to implement the environmental priorities stated in the policy
- Perform the activities necessary to achieve the objectives and targets, develop documents and records, and train employees in their environmental responsibilities
- Monitor and measure on a regular basis the performance of the environmental management system
- Review the entire set of environmental management activities periodically to ensure continual improvement

These five basic activities are intended to facilitate the allocation of resources; the assignment of responsibilities; and the ongoing evaluation of practices, procedures, and processes to ensure achievement of the environmental policy requirements. Thus, the ISO 14001 standard views the business from the process perspective; defines the boundaries of environmental responsibility through the policy statement; and seeks structural solutions through comprehensive monitoring, internal audits, and management reviews.

ISO 14001 recognizes the importance of governments, industry leaders, and stakeholders working together to design and implement administrative and structural procedures to improve environmental conditions. Most businesses will have to modify their corporate behavior to implement the standard. In particular, some will have to learn how to develop better working relationships with the public and engage in collaborative decision making to improve environmental performance. Governments will need to redirect regulatory agencies to change existing regulatory schemes and create further incentives for business to protect the environment. The federal and state governments will need to develop better working relationships with each other so that they no longer compete and interfere with each other's statutory responsibilities, as the manner in which businesses self-regulate and report to these entities changes.

Skeptics may find the tasks required to effect these global changes overwhelming. The evolution of these changes, however, is already beginning. Later in this book we address some of these initiatives and set forth how businesses can take advantage of federal and state programs designed to allow businesses to self-disclose environmental problems and take effective corrective action. These programs can often significantly reduce or eliminate penalties for offenses that occur. The U.S. Department of Energy (DOE), under its General Environmental Protection Program, is adopting the ISO 14000 approach and developing its own environmental management systems. Throughout Europe and Asia, government bodies are working with industry leaders to implement the requirements of ISO 14001.

ISO 14001 provides businesses with a practical and workable framework for managing environmental risk. Its focus on continual improvement and prevention of pollution encourages businesses to move from risk financing into comprehensive risk management activities. It also ensures that a business adopts the process perspective and systematically evaluates and analyzes existing and potential exposures before losses occur. With ISO 14001, environmental decisions become strategic concerns, which enables risk assessment and long-term resource allocation—while considering both external pressures and operational priorities.

Implementation of a comprehensive and integrated environmental management system has the effect of simultaneously managing and reducing environmental risk. These two separate but interrelated concepts are merged in the systems approach. ISO 14001 requires businesses to determine the legal and other requirements and environmental aspects associated with their activities, products, and services. A process must be defined by businesses to achieve target performance levels and engage in environmental planning throughout the product or process life cycle. Companies are required to provide appropriate and sufficient resources to achieve targeted enforcement levels on an ongoing basis, while establishing and maintaining communications with internal and external interested parties. Appropriate and sufficient resources must also be allocated for training to achieve targeted performance levels on an ongoing basis. Finally, ISO 14001 provides management with a process to audit and review the environmental management system and to identify opportunities for improvement of the system and the resulting environmental performance.

The ISO 14000 environmental management system, coupled with other risk management and risk reduction techniques, comprises the systems approach. In various sections throughout this book, we explore in more detail how ISO 14000 works and provide an overview for businesses to begin to consider how the standard can be applied to their systems operations.

Chapter 1 introduced the nine basic building blocks of an effective EMS, reflected in ISO 14001 and other models. They include (1) top management leadership, policy setting, and review; (2) identifying environmental issues/ risks (or "aspects," in the vocabulary of ISO 14001) and legal and other requirements; (3) establishing objectives to successfully manage these risks and requirements in line with policy; (4) implementing programs and procedures (including for emergencies) that establish how, by whom, and when those objectives will be met; (5) training people so that they know their risks and what to do to best manage them; (6) monitoring, measuring, and auditing to track performance and verify implementation; (7) taking preventative and corrective action; (8) continual improvement; and (9) documentation and record keeping (Bell, 2018). The ISO 14001: 2015 standard itself defines these building blocks under the following headings: Leadership, Planning, Support, Operation, Performance Evaluation, and Improvement. Within those organizational activities are various environmental management functions, indicated in the box below. This chapter will illustrate management activities within the framework of ISO 14001 that are organized and implemented to create an effective environmental management system.

ISO 14001: 2015 ENVIRONMENTAL MANAGEMENT SYSTEMS—REQUIREMENTS WITH GUIDANCE FOR USE

CONTEXT OF THE ORGANIZATION

4.1 Understanding the organization and its context

4.2 Understanding the needs and expectations of interested parties

4.3 Determining the scope of the environmental management system

4.4 Environmental management systems

LEADERSHIP

5.1 Leadership and commitment

5.2 Environmental policy

5.3 Organizational roles, responsibilities and authorities

PLANNING

6.1 Actions to address risks and opportunities

6.1.1 General

6.1.2 Environmental aspects

6.1.3 Compliance obligations

6.1.4 Planning action

2.3 Environmental Policy

Since ISO 14001's initial publication in 1996, the environmental policy statement has been an important driver of each business's environmental management system. In the ISO 14001 process, the environmental policy identifies objectives and targets and establishes an environmental program as the backbone of the risk management plan. The policy provides the impetus for a series of actions to minimize risks emanating from uncontrolled emissions, noncompliance with regulatory requirements, inadequate environmental management practices, and uncorrected adverse conditions. It also provides the opportunity to further reduce environmental risks by removing or minimizing the causes of pollution and any remaining liabilities from previous incidents.

ISO 14001 requires top management to develop a corporate environmental policy statement and to ensure that it

- Is appropriate for the nature, scale, and environmental impacts of the activities, products, or services of the company
- Includes a commitment to continual improvement and prevention of pollution
- Includes a commitment to comply with relevant environmental legislation and regulations, and with other requirements to which the organization subscribes
- Provides the framework for setting and reviewing environmental objectives and targets
- Is documented, implemented, maintained, and communicated to all employees
- Is available to the public

The corporate environmental policy statement is the driver for implementing a new environmental management system. Both the policy and the system then can be used by the company to maintain and improve its environmental performance. The policy establishes an overall sense of direction and sets the principles of action for the organization. Later in this book, we will explore environmental policy statements and discuss the importance of a strong corporate environmental policy statement.

2.4 ISO 14001:2015

As noted, the purpose of this book is not to instruct the reader on how to design and implement ISO 14000 systems, and the reader is encouraged to purchase a copy of the ISO 14001:2015 standard from their national standards body and to read the many excellent books available on the topic to gain a more detailed understanding of the ISO 14000 standards. Since ISO 14001 plays such an outsized role in the implementation of most EMSs, the following discussions are intended to bring the reader up to date on the significant modifications made to ISO 14001 in the 2015 revision.

ISO 14001:2015 was the first major revision to the standard since its initial publication in 1996. The most significant change was to bring the standard in line with ISO's generic management system framework, known as Annex SL, the purpose of which is to create consistency among the common elements of ISO's various management systems standards. In addition to normalizing the language, structure, concepts, and vocabulary, the revised standard benefited from making use of the many implementation practices developed over almost 20 years in use and from a desire to increase the emphasis on taking a more strategic approach to environmental management.

As detailed in Bell (2018), the ISO 14001:2015 revision maintains the "plan-do-check-act" framework of the original standards. However, the structure and phraseology were changed to orient the "planning," "support," "performance," and "improvement" around the focus on leadership. The order of the various clauses was changed to encourage organizations to have an overall strategic understanding of their place and purpose, including expectations of interested parties.

The revised standard adds a new Clause 5 titled "Leadership" that presents not only issues of the corporate policy and the management review but also more detail on top management's obligations, structure, and responsibility. Consistent with the general goal of more explicitly linking the EMS with the bigger picture, top management is obliged to integrate the EMS into the organization's business processes and make certain that the environmental policy and objectives are compatible with the strategic direction of the organization.

The planning elements in Clause 6 remain largely the same, but the clause adds a significant amount of detail presenting the common practices developed over the last several decades of standard use. More attention is paid to strategic planning, including issues of life-cycle perspectives, and to not only knowing legal obligations but also determining how the legal obligations apply to the organization.

ISO 14001:2015 contains many more references to compliance, though the treatment of "compliance obligations" may concern many in the United States. A requirement is defined as a "need or expectation that is stated, generally implied or obligatory," and organizations are expected to determine which "requirements" of "interested parties" must become "compliance obligations." Particularly in the United States, one should be very cautious in redefining such a central concept as "compliance obligations," which brings with it an array of long-established legal, risk management, and other liability consequences. Trying to explain that an organization requirement voluntarily entered into is a compliance obligation may create both internal and external risk management and communications challenges (Bell 2018).

Based on a common response to the user community's frustration over the volume of documentation generated by implementing ISO management systems standards, there is a decrease in the number of places in the standard that documentation is required. ISO 14001:2015 is accompanied by a 13-page informative annex that is intended to provide guidance, but it does not establish new or different requirements and does not contain auditable criteria.

Overall, the current ISO 14001:2015 standard reflects its drafters' efforts to harmonize the EMS standard with other ISO management systems standards as well as their desire to add more detail reflecting several decades of implementation experience. ISO 14001:2015 takes a more strategic approach to designing and implementing EMSs, taking into account broad themes such as sustainable development, life-cycle thinking, transparency, and stakeholder engagement.

The ISO 14001 standard can serve as an excellent organizational risk management framework in which a business considers, plans for, and manages environmental aspects and impacts before extreme conditions exist. ISO 14001 is widely recognized by the U.S. EPA and many state environmental regulators as one of the best EMS models; in fact, the EPA uses it for its own operations. By setting up systems that conform to the standard, senior management will be confident that it has handled its environmental risks and limited its exposures in a systematic and controlled manner. In ISO 14001, environmental decisions are strategic—risks can be assessed and long-term resource allocations can be made in light of competitive considerations and existing operational priorities.

2.4.1 Assessing Risks

Management often receives conflicting advice regarding the risks associated with operations and properties. It is thus important that managers have an adequate framework to accurately assess and prioritize environmental risks. ISO 14000 provides a structure to manage and understand various environmental risks facing a business.

Environmental risk traditionally has included historic liability risk, operating risk, marketplace risk, capital cost risk, transaction risk, and sustainability. Although the historic liability risks identified during an environmental site assessment are increasingly quantifiable, little hard data are available regarding the assessment and management of ongoing operational risk.

Environmental risk quantification differs from quantification of other casualty risks because of the relatively small amount of historical data available and the complexity of the contamination distribution variables. Ten years of worker's compensation loss experience, for example, is useful information upon which to base future loss projections, but ten years of soil contamination data are not likely to accurately predict future liabilities at a given site. Numerous variables, such as tank construction and installation, stored contents, local hydrogeology and geology, surrounding populations, and applicable laws and regulations, make it difficult to predict future environmental outcomes. In the absence of adequate historical data, the management systems, or engineering controls, must be used to develop an accurate assessment of environmental risk.

2.4.2 Risk Modeling

Experts in environmental, insurance, and risk management firms have developed risk modeling techniques to be used in the ongoing assessment of environmental risk. After the business has performed the initial review of its environmental aspects and impacts, legal and other requirements, and has obtained its risk baseline, those data can be used in risk modeling to plan and prioritize.

Generally, risk modeling is a method of breaking down a complex situation into its component parts, arranging those parts in hierarchical order, assigning numerical values to subjective judgments regarding the relative importance of each variable, and synthesizing the judgments to determine which variables have the highest priority. By using these models effectively, a business can act upon the highest risk concerns to predict and prevent loss, thereby controlling risk.

ISO 14000 can utilize the results from these risk assessing and risk modeling processes in developing a comprehensive environmental management system that properly manages and ultimately reduces these risks. The benefits of developing a systems approach using an ISO 14001–compliant environmental management system as its centerpiece are set forth in the next chapter.

References

Anonymous. "Report Finds Air Act's Benefits May be 70 Times Higher than Its Costs." *Environment Reporter*, November 15, 1996.

Bell, C.L. "ISO 14001 and Environmental Management Systems: Where Are We?" *A&WMA, The Magazine for Environmental Managers*, July 2018.

Boiral, O., L. Guillaumie, I. Heras-Saizarbitoria, and C.V.T. Tene. "Adoption and Outcomes of ISO 14001: A Systematic Review." *International Journal of Management Reviews* 20, no. 2 (2018): 411–32.

Campos, L.M.S., D.A. de Melo Heizen, M.A. Verdinelli, and P.A.C. Miguel. "Environmental Performance Indicators: A Study on ISO 14001 Certified Companies." *Journal of Cleaner Production* 99 (July 15, 2015): 286–96.

Heras-Saizarbitoria, I., J.F. Molina-Azorín, and G.P.M. Dick. "ISO 14001 Certification and Financial Performance: Selection-Effect Versus Treatment-Effect." *Journal of Cleaner Production* 19, no. 1 (January, 2011): 1–12.

ISO. "ISO 14000 Family—Environmental Management." https://www.iso.org/iso-1 4001-environmental-management.html.

ISO. *The ISO Survey of Management System Standards Certifications 2016*, 2017. https://www.iso.org/.

Ratnasingam, J., K. Wagner, and S.R. Albakshi. "The Impact of ISO 14001 on Production Management Practices: A Survey of Malaysian Wooden Furniture Manufacturers." *Journal of Applied Sciences* 9, no. 22 (2009): 4081–85.

Reis, A.V., et al. "Is ISO 14001 Certification Really Good to the Company? A Critical Analysis." *Production* 28 (2018): e20180073. doi:10.1590/0103-6513.20180073.

Riaz, H., A. Saeed, M.S. Baloch, and Z.A. Khan. "Valuation of Environmental Management Standard ISO 14001: Evidence from an Emerging Market." *Journal of Risk and Financial Management* 12, no. 1 (2019): 21.

Tarí, J.J., J.F. Molina-Azorín, and I. Heras-Saizarbitoria. "Benefits of the ISO 9001 and ISO 14001 Standards: A Literature Review." *Journal of Industrial Engineering and Management* 5, no. 2 (2012): 296–322.

US EPA. "Benefits and Costs of the Clean Air Act 1990–2020, The Second Prospective Study." 2011. https://www.epa.gov/clean-air-act-overview/benefits-and-costs-clean-air-act-1990-2020-second-prospective-study.

US Government Accounting Office. "Superfund: Information on the Nature and Costs of Cleanup Activities at Three Landfills in the Gulf Coast Region. GAO-11-287R." 2011. https://www.gao.gov/products/GAO-11-287R.

Vilchez, V.F. "The Dark Side of ISO 14001: The Symbolic Environmental Behaviour." *European Research on Management and Business Economics* 23, no. 1 (January–April, 2017): 33–39.

Zobel, T. "The Impact of ISO 14001 on Corporate Environmental Performance: A Study of Swedish Manufacturing Firms." *Journal of Environmental Planning and Management* 59, no. 4 (2015): 1–20.

For a perspective of the U.S. government on compliance systems generally, see §8B2.1 of the U.S. Sentencing Commission's Sentencing Guidelines Manual, which describes the elements of an ethics and compliance program, and Principles of Federal Prosecution of Business Organizations, 9–28.000 of the U.S. Attorney's Manual. For a selection of EPA's views on EMSes, see www.epa.gov/ems; Compliance-Focused Environmental Management Systems–Enforcement Agreement Guidance (EPA Jan. 2005); and Guidance on the use of Environmental Management Systems in Enforcement Settlements as Injunctive Relief and Supplemental Environmental Projects (EPA June 2003).

For a more detailed discussion of EMSs and environmental law, see Bell, C. "Environmental Management Systems and Environmental Law." In *Environmental Law Handbook.* 23rd ed. 1057–1100. Plymouth, UK: Bernan Press, 2017.

3

Preventing and Mitigating Environmental Liabilities with Environmental Management Systems

Robert A. Woellner

Management should consider the benefits of developing and implementing an environmental management system from six perspectives: economic, social, political, technological, ideological, and financial. Opportunities and risks can arise in a business in any or all of these areas. By reviewing each perspective, the problem is viewed in a macro context, wherein meaningful evaluations of the business's operations and management systems can be made.

3.1 Economic Benefits

The economic sector consists of the global, national, and local conditions of production, distribution, and service. Included are consumer spending, inflation rates, interest rates, labor supply, cost and availability of natural resources, financing methods, industry performance ratios, business investment patterns, and other conditions of production and competition.

Adopting an ISO 14000–type environmental management system may assist a business in achieving a qualified-supplier status, which can result in an expanded customer base. As numerous businesses obtain their ISO 14001 certifications, increasing pressure will be placed on third- and fourth-level suppliers to become certified.

Internal operating efficiencies and cost reductions are also benefits to the business. By performing a process analysis and defining in detail operating processes that impact the environment, management also has the opportunity to reduce duplicative efforts and eliminate redundant systems. Very often the exercise of process mapping reveals many convoluted processes that have developed over time without being rationalized, evaluated, or balanced with other similar processes.

Process redesign, which identifies and isolates polluting processes, can assist management in reducing energy usage and waste. Prevention of pollution, waste minimization, substitution of materials with less toxic ones, and designing to reduce environmental impacts are some techniques management can use to reengineer existing processes to reduce or eliminate waste and polluting by-products. Incorporating a life-cycle perspective allows management to extend its control over environmental impacts beyond the facility and to address issues arising from the types of raw material streams used, from the transportation of raw and processed materials, and from ineffective product disposal. These issues, when properly addressed, can translate into operating efficiencies and cost reduction regardless of the life-cycle stage in which the business operates. The life-cycle perspective also allows management to look at the possibility of recovering costs through recycling of packaging and manufacturing wastes, as well as through facility operations, upstream operations such as raw material extraction, and downstream operations such as product use.

A major challenge for senior management is to decrease liability exposure from environmental incidents and accidents. By identifying high-risk processes and working to reduce their environmental impacts, management can address areas of concern and ensure that risks are systematically reduced or eliminated. These efforts should be detailed in an environmental management system so that the business can obtain cost reductions through reduced insurance rates and access to capital at below-market rates. The Federal Deposit Insurance Corporation guidelines for lenders prescribe activities that evaluate the environmental management practices of potential borrowers. The Environmental Bankers Association has developed a clearinghouse for information on the environmental practices and procedures of the lending industry sourced from the government and its member institutions. By implementing a formal environmental management system, a business can demonstrate to lenders that it meets or exceeds accepted lending standards, thus ensuring access to capital and maintaining positive relations with lending institutions.

As discussed in subsequent chapters, numerous court decisions consider evidence of a good environmental management system for purposes of penalty mitigation. According to the *Federal Sentencing Guidelines for Organizations*, fines imposed on a business for violating environmental laws have a wide potential range and can be reduced based on six culpability criteria: the level of authority and size of the organization; prior history; violation of an order; obstruction of justice; existence of an effective program to prevent and detect violations of law; and self-reporting, cooperation, and acceptance of responsibility. The guidelines allow for either a reduction of Environmental Protection Agency (EPA) penalties for a business with an environmental management system in place or an increase in EPA penalties for a business that does not demonstrate responsible environmental management. The implementation of a formal environmental management system addresses several of these responsible management criteria to allow for mitigation of fines should a violation occur.

Large businesses are frequently faced with a number of different customer, regulatory, and registrar audits of their operations. Those doing business internationally are also often required to perform multiple inspections, certifications, and product registrations in order to demonstrate conformance with a varying array of regulations, requirements, and other technical specifications. The activities required under an ISO 14001–type environmental management system provide consistent performance data and rationalized processes that may allow management to reduce the number of environmental impacts and the systems required to respond to audits and inspections.

3.2 Social Benefits

The social arena is focused on people, communities, and society at large. It includes population characteristics and trends, lifestyles, values, ethical standards, attitudes, public opinion, educational patterns, social change movements, and nonprofit groups and organizations. When a business adopts an environmental management system, the social perspective allows management to see how well the system can integrate the business into society at large.

Over the past few decades, the growing awareness of the environment as an important issue has led to the emergence of opportunities for businesses with environmental impacts. A business can project a socially responsible image by integrating an environmental management system into its operations. To ensure that the business's communication efforts are not mistaken for *greenwashing*, the use of national and international standards can lend much credibility to the business's claims. These standards provide a consensus approach that has been debated and agreed upon by experts throughout the world. Adoption of standards such as ISO 14001 enables a business to demonstrate a sincere and credible commitment to the environment, and to base its claims on a system that represents the state of the art worldwide.

As business leaders begin to broaden their perspectives, they realize that a wide array of stakeholders or interested parties are affected by and concerned with their operations. The growing trends of environmental activism, heightened awareness of environmental impacts due to increased reporting, and decreasing acceptance of businesses seen as environmental bad actors provide an opportunity for businesses to satisfy stakeholder interests for corporate accountability. A business can promote environmental awareness and ecological responsibility as cornerstones of its operating principles by developing and implementing a good corporate citizen policy as the basis for its environmental management system. By using a communications-based framework to demonstrate its commitment to these principles, management can embed environmental excellence into formal and informal reports, town meetings, and product and services descriptions.

A key aspect of any ISO 14001 environmental management system is the integration of environmental issues into strategic decisions. In fact, ISO notes that "increased prominence of environmental management within the organization's strategic planning processes" is one of the main updates made to the 14001 standard in its 2015 revision. By proactively addressing existing trends and emerging societal issues, a business can ensure that stakeholder perceptions of the risks posed by the business are accurate. Through continual strategic reviewing of the environmental management system, management can be assured that it is monitoring and addressing issues and trends that could have an adverse impact on the business. Proactive management also ensures that management is holding itself accountable to the wide variety of stakeholder interests and concerns.

Customer relations can be positively affected by the presence of a formal environmental management system. Customers can gain assurance that the business is proactively managing its environmental activities to minimize the likelihood of environmental incidents or accidents. Customers, end producers, and contractors, all gain more confidence in the integrity of management and are more likely to continue and extend their relationships when a business takes steps to ensure an uninterrupted supply of products and services.

How a business is performing from its customers' perspective has become a priority for well-run businesses. It is imperative that a business translate its general mission statement on customer service into specific environmental objectives and measures that reflect the concerns truly important to customers—cost, performance, quality, service, and time.

Although management may enact controls for minimizing the release of a certain chemical, the unintended use or disposal of products may have a severe effect on customer perception of business as being environmentally responsible. During the management review, senior management should address the implicit and explicit aspects of this issue to take advantage of the opportunity to build customer satisfaction. Management systems can identify the financial reasons and environmental consequences of product labeling, product redesign, awareness and elimination of toxic materials, and packaging or development of infrastructure for product reuse and recycling. Innovations made possible by top management's commitment to environmental concerns can save resources and can have a substantial impact on the environment. Later chapters discuss how environmental management systems have caused innovations that have improved environmental performance.

3.3 Political Benefits

The current permitting and reporting processes required by environmental regulations can be time-consuming and difficult. An ISO 14001 environmental management system provides a means for consolidating and normalizing environmental documents and data, thus allowing faster permitting and reporting processes to be implemented. The rationalization of processes will usually be accompanied by an increase in data accuracy and lend greater integrity to the permits and reports submitted to regulatory authorities. In recognition of this integrity, regulators will be much less likely to seek enforcement actions and will increasingly rely on a business's internal management should a violation occur.

A key area of political benefit arises from the self-assessment and self-management of the technical issues associated with the environmental aspects of operations. Frequently, a business expends resources to comply with environmental regulations, but no significant environmental benefit is realized by that business from those expenditures. In some cases, the technical approaches fostered by the regulations cause the business to focus on areas of activities that have minimal environmental impact, while other areas with severe impacts are left relatively uncontrolled. The EPA has recognized this dilemma by stating that compliance with regulations and improved environmental performance are not necessarily the same thing. One-size-fits-all regulations that prescribe what to do and how to do it can impede technical improvements that would otherwise lead to improving the environment.

Through both the enhanced communication and the increased reporting integrity fostered by an ISO 14001 environmental management system, a business can address the technical management of processes unique to its industry and work with regulators to prevent ineffective and costly command-and-control initiatives from becoming regulations. The opportunity lies in moving the regulators away from a micromanagement approach, which is best controlled by the people closest to the processes, to a macromanagement approach. Industry needs to be allowed to seek its own solutions while continuing to meet specific performance levels required by law.

An ISO 14001 environmental management system can provide a common, systematic approach for international businesses working in countries with different regulatory frameworks. It is a practical means to proactively manage regulatory compliance, regardless of the content of those regulations. By adopting a simple, pragmatic method of accessing regulatory requirements and ensuring compliance with these requirements, the ISO 14001 framework encourages effective environmental management within a wide range of regulations and laws.

3.4 Technological Benefits

As a management system, ISO 14001 can be considered a technology in itself. Although the focus of the standard is management systems, these systems are intended to improve environmental performance by preventing pollution and removing systemic causes of noncompliance. The systems approach looks at the interrelationships of many activities and how they work together to achieve implementation of the environmental policy. By analyzing processes for their environmental impacts, management obtains a clear picture of how the business functions with respect to the environment. Management can then make better decisions about allocating scarce resources to minimize or eliminate negative impacts.

When management focuses on aligning processes with stakeholder requirements for environmental performance, it reveals both performance gaps and improvement opportunities. Decisions about environmental technologies should be internalized, so that management can easily exceed the regulatory requirements and integrate environmental issues into business decisions. Both of these elements provide the foundation for continual improvement of environmental performance and foster the integration of technical environmental activities and overall business strategies.

A central feature of an ISO 14000–type environmental management system is that it provides a single environmental management system for global organizations or businesses with multiple sites and facilities. Consider the example of financial management systems. How would most businesses function if they paid their salaries and expenses at different times using different forms of currency? What if they reported their activities irregularly, incompletely, and not according to a standard format? Most businesses would disintegrate into chaos with such a fragmented financial management system. By using a standard environmental management system, the business can define and control its environmental impacts and track them in ways familiar to financial managers.

3.5 Ideological Benefits

Ideology is characterized by the ideas and concepts that societies throughout the world embrace as interpretations of reality. These ideas include religion, science, philosophy, and the arts. Emerging ideas and concepts in these areas have foretold massive shifts in understanding, such as the Renaissance, the Industrial Revolution, and now the Knowledge Revolution.

Beginning with the first Earth Day in 1970, and more recently the Paris Climate Change Conference in 2015, an emerging awareness of environmental responsibility has been taking hold in the industrialized and developing nations of the world. This multifaceted idea consists of several concepts: sustainable development, ecological integration, intergenerational responsibility, and natural resource stewardship. People throughout the world are searching for solutions to environmental problems involving the contamination of air, water, and soil. They are uniformly rejecting the throwaway consumer mentality of the 20th century. The synchronism of their ideas demonstrates that universal environmental reform is what people want and that the same solutions can be applied to ecological problems throughout the world.

ISO 14000 represents a series of methods to transfer these ideas and concepts to the business context. The usable management systems of ISO 14000 can be incorporated into operations and can provide a bridge between the heavily polluting, resource-intensive industries that characterized the early phases of the Industrial Revolution and the new, cleaner technologies of the 21st century. Businesses that understand and embrace the concepts of sustainable development, ecological integration, intergenerational responsibility, and stewardship will find ISO 14001 to be an indispensable means to achieving these important goals. Increasingly, wealth in the future will be created from managed information, not from depleting natural resources or manufacturing.

3.6 Financial Benefits

In an ideal world, all people and businesses would manage their environmental affairs purely for moral, ethical, and social reasons. The public is now beginning to realize, however, that the best way to achieve environmental change is to reward effective, proactive environmental management. Positive incentives, as well as clear direction, are crucial for the attainment of environmental goals. Financial reward is an acceptable and appropriate motive for environmental improvement.

Financial benefits of effective environmental management can be attained by increasing consumer and shareholder confidence; reducing the costs of doing business; improving relationships with investment bankers, commercial lenders, private equity, and the stock and bond brokerage community; boosting management and employee morale; increasing profits; and cutting legal and administrative costs. These benefits can have a substantial and long-lasting effect on the financial viability of a business.

The ISO 14001 standard for environmental management provides a practical and workable framework for controlling environmental risk. Its focus on continual improvement and pollution prevention encourages businesses to move from reactive risk management and risk financing into comprehensive risk-control activities. Following this path, businesses assure themselves and stakeholders, including financial partners, that they are identifying, prioritizing, and actively managing environmental exposures to lessen the likelihood of loss. Financial stability is an important by-product of a well-managed environmental system.

3.7 Costs of Implementation

As with any investment, management needs to be aware of the operational and financial consequences that the development and implementation of an environmental management system will have on the business. One of the benefits of implementation can be the identification of costs associated with environmental activities, thus enabling management to make more logical decisions about pollution prevention and waste minimization alternatives.

The first major category of implementation costs is internal resources. These costs usually represent the majority of the expenditures and include personnel time, training, and information technology to support the new flow of environmental information. Personnel will be involved in developing documentation, defining process and information needs, and managing the project and its myriad activities. Depending upon the results of the initial review and process analysis, more measuring and monitoring equipment or upgrades to current equipment may be required. Equipment costs can be capitalized and amortized, while personnel time is generally an operating expense in the period during which it occurred. Internal resources account for approximately 80% of the costs of implementing an environmental management system.

Although the majority of costs usually relate to internal resources, many businesses find that they do not possess the required expertise to fully develop the required systems. A business may use external consultants and purchase commercially available training programs for its employees. A business may find that investments in outside expertise can reduce implementation time considerably by providing critical assistance to the staff who will be tasked with developing and implementing the environmental management system.

Should a business seek registration under the ISO 14001 standard, it will require the services of a registrar. These services are both one-time and continuing. One-time costs include the initial pre-assessment and registration audits. The business also needs to demonstrate continual conformance with the standard, since recertification is required under the current accreditation scheme, so management can anticipate periodic conformance audits performed by the registrar as a cost of maintaining registration.

Virtually every business faces the possibility of environmental liability costs. Costs may be derived from lawsuits involving customers, employees, or communities or from legally mandated cleanup of hazardous waste sites. It is essential that senior managers make at least a general estimate of their business's potential future environmental liability. The U.S. Securities and Exchange Commission (SEC) requires adequate disclosure of environmental matters with it (Armao and Griffith 1997; Kuhnhein and Burke 2010). A logical and accurate assessment of potential liabilities will not only allow for the allocation of suitable levels of financial resources but also articulate a comprehensive risk management program and a reassessment of corporate strategy and management practices.

The initial development of an environmental management system can be time-consuming and expensive, but it typically costs a fraction of the potential litigation and cleanup costs. If an environmental management system is properly implemented, the investment should result in impressive annual savings derived from improved management efficiencies.

From a financial perspective, a business should determine how its environmental performance is viewed by shareholders, lenders, insurers, and employees. Financial risks include loss of profit, loss of access to capital, loss of market share, and loss of business. Benefits include an enhanced ability to survive, succeed, and prosper.

By implementing an environmental management system, senior management ensures that it is effectively handling its environmental risks and limiting its exposures, addressing structural solutions, and reducing the need for excessive risk financing.

3.7.1 Cost/Benefit Analysis

As part of the process analysis, management must analyze various environmental costs, including hidden expenses, in reaching better decisions regarding systems operations. The key to evaluating the investment in an environmental management system is to derive a set of costs that accurately reflects current processes and match it with a set of benefits incurred as a result of the environmental management system. In order to do this, management should attempt to match the temporal, quantitative, and qualitative characteristics of costs and benefits to more clearly understand the relationship between them.

The full range of environmental costs for an environmental management system will include both implementation and operating costs. Once the initial expenditure for system development is made, ongoing costs would include the following:

- Payroll and direct personnel expenses associated with system activities
- Environmental management system and compliance audits

- Environmental testing
- Purchase, calibration, and operation of environmental monitoring equipment
- Environmental cleanup costs
- Environmental liability insurance premiums
- Waste disposal costs
- Reserves set aside for contingent liabilities

The key to performing an effective cost-benefit analysis is the ability to calculate the specific environment-related costs associated with producing a particular product or service. Analyzing the costs and the benefits realized from implementing the environmental policy and strategy can answer two basic business questions for management:

- Have the environmental activities resulted in an increase in sales?
- Have the environmental activities resulted in a decrease in operating costs?

The environmental management system must demonstrate that it contributes positively to the operations and the financial status of the business to be accepted as a reasonable investment. A typical presentation to management will show one-time investments resulting in long-term lower operating costs. Standard financial evaluation techniques, such as the internal rate of return or return on investment, can be used. Indirect links to increased sales could also be shown to further favor the investment.

The problem with evaluating the benefits of an environmental management system is that even though the costs incurred will be substantial and immediate, many of the benefits will be long-term and not completely quantifiable. Many direct cost reductions in waste disposal and energy and material usage could be realized, while other environmental management system activities could be placed under the rubric of risk prevention. In evaluating the financial performance of risk-prevention activities, management will have to look at the opportunity costs or avoided costs of preventive activities in their analysis. That these opportunities and avoided costs can be difficult to estimate should not deter management from making a basic evaluation of their magnitude.

Many businesses have redesigned their processes to eliminate toxic chemical use, thus avoiding the direct costs of disposal and pollution control equipment altogether. More difficult to estimate, however, is the amount of fines, penalties, legal fees, cleanup expenses, and revenue loss because of negative public relations that the business would have incurred had it continued to operate by utilizing the old procedures. In this type of analysis, it is important to remember the benefits to stakeholder relationships, including increased customer acceptance and loyalty, a business can gain by communicating the results of its environmental activities and maintaining a favorable corporate image.

3.7.2 Domestic and International Industry-Specific Certification Advantages

Certification can benefit specific industry sectors, both in the U.S. and internationally. Each industry sector is different and faces unique environmental challenges. Some of the industry groups that benefit the most from obtaining certification are general manufacturing, automotive manufacturing, chemical manufacturing, electronic equipment and component manufacturing, service industries, and health and pharmaceutical industries.

3.7.3 International Advantages

When considered from an international perspective, the advantages of ISO 14001 certification change considerably. Within the industrialized nations, the overriding benefit becomes the ability to reduce pollution and maintain social credibility. The already developed structures for certification provide an important recognition status for ISO 14001 implementation. When considered from the macro perspective, however, the main benefit of an ISO 14001 environmental management system is the demonstration of responsible environmental management. Management can also be assured through process efficiencies, reduction of waste, and proactive communications that the business is perceived as a good neighbor and a preferred supplier by its customers, communities, and markets.

In the developing countries, ISO 14000 certifications are considered an asset to international competitiveness. Businesses located in these countries need to secure a leading-edge position with regard to modern technology. Obtaining ISO 14001 registration encourages advancement in the world marketplace and removes stigmas associated with outdated environmental management practices. Export markets are crucial to the success of emerging nations, especially for the extractive and manufacturing industries. These businesses can use the management technology presented in the ISO 14000 standards to advance not only their environmental practices, but also their overall strategic and operational management planning.

As the ISO 14001 environmental management system standard replaces the plethora of national and regional environmental management standards worldwide, industries in developing countries will realize a distinct trade advantage by obtaining certification. The fewer standards with which a business must comply, the easier it will be to sell its products across the world. As industries in developing countries grow, they also have the opportunity to build environmental management systems into their structure from the very beginning, thus obtaining a cost advantage that increases with time.

Adoption of the ISO 14001 standard by developing countries can ensure that the push for improved corporate environmental quality does not become a hindrance to international trade.

The benefits of developing a system to manage and reduce environmental risks will inevitably outweigh the costs. The advantages can include increased productivity and morale as a result of joint enterprises that positively impact the environment. As discussed in the next chapter, these positive impacts can also include the development of a single integrated strategic information system that can be understood and fully utilized by management, employees, and customers.

One of the most important components of any environmental risk management strategy is the definition and control of organizational communications methods. Regardless of the actual risks posed by a business's activities, it is the stakeholder perceptions of these risks that can determine the success or failure of the environmental management system. These risk perceptions, which drive community and public interest group activities, can only be managed effectively by a clearly defined set of communications principles, procedures, and goals. Thus, management must develop and implement a comprehensive communications strategy aligned with the business's environmental policy, impacts, and regulatory requirements.

Effective environmental communications strategies have two primary benefits. First, they provide the opportunity to reduce duplicative data collection, reporting, and record keeping. Effective communications also benefit the overall strategic positioning of the firm. By evaluating and rationalizing its information needs, the business establishes the basis for implementing an automated information management technology. An environmental communications strategy can provide much-needed synergy between an information technology initiative and environmental management activities.

3.8 Defining the Boundaries of the Communications Strategy

Businesses can select two basic forms of communications. Participatory communications require the active involvement of the audience and establish a two-way system that allows the sharing of ideas. Interactive communications need facilitation and active solicitation of audience feedback to ensure that a common understanding is developed and comments are given adequate consideration.

The second form of communication, nonparticipatory, is a one-way message delivery to an intended audience. It is generally limited to providing information while the audience remains passive. Nonparticipatory communications do not ensure that a common understanding is achieved but may be appropriate for many situations in which the business is responding to requests for information or is providing routine information required by regulatory agencies.

The ISO 14001 standard requires two methods of communications: internal to the business (management and employees) and external to the stakeholders. Both forms of communication should be founded on a common set of facts and ideas to ensure consistency, accuracy, and completeness. A functioning environmental management system can provide the data upon which the internal and external methods are built. Once these program elements have been defined, the business can develop and implement specific communication methods based on their effectiveness with intended audiences.

ISO 14001:2015 requires that the organization identify which interested parties are relevant to the EMS, the relevant needs and expectations of those interested parties, and which of these needs and expectations must become its compliance obligations, and then take these compliance obligations into account when developing and implementing its environmental management system. The management system must take into account what, when, with whom, and how to communicate with interested parties. The EMS communications must include both internal communications that are relevant to the EMS throughout the organization, to ensure that the persons doing work under the organization's control contribute to continual improvement, and external communication of information relevant to the EMS as required by the organization's communication processes and as required by its compliance obligations.

3.9 Internal Risk Communications

Internal risk communications are directed toward employees and stakeholders and should be focused on achieving three objectives:

- Demonstrating management commitment to responsible environmental management
- Responding to questions and concerns about the business's environmental management activities
- Increasing awareness about the business's environmental policy, objectives, targets, and employee environmental responsibilities

Businesses should establish an internal environmental risk communications program based on their overall internal communications policy, as well as the environmental policy. This communications policy can usually be found in an employee handbook or a compilation of company policies. The contents and activities of the internal communications program can address employee concerns and needs, the environmental aspects and impacts of the business, and the ongoing environmental activities.

Establishing a set of roles, responsibilities, and accountabilities for environmental performance provides the foundation for internal communication activities and drives the type and frequency of individual communication events. Management can ascertain employee concerns and needs about environmental management by using a survey, the results of which can also be incorporated into the management review process of the environmental management system.

Management must also identify the scope of internal communications activities with regard to significant environmental aspects of business operations. While there are regulations defining the external reporting requirements for an environmental incident, the business must define exactly what comprises an incident for internal communication purposes. While not every incident may need to be communicated to employees, the results of incident investigations can play an important part in formulating corrective and preventive actions, as well as in the management review. It is important to remember that incidents may already be the subject of rumors and the informal communication network. Management should therefore take a proactive role to ensure that accurate and timely information is distributed to employees and other stakeholders.

3.9.1 Internal Communications Methods

Internal communications methods can vary widely according to the size, type, and environmental aspects and impacts of the business. Some common methods include the following:

- Daily, weekly, or monthly meetings
- Newsletters
- Pay envelope messages
- Notices in the workplace
- Websites
- Social media
- Video conferences
- Training sessions, on-site and off-site
- Performance reviews
- Intranet applications (e-mail)

3.10 External Risk Communications

External environmental risk communications involve the development and presentation of an accurate account of the effective management of the company's risks. Widely varying groups of stakeholders need to understand how environmental risk is managed within the company and not by or as a result of outside influences. Credibility, timeliness, appropriateness, and completeness become key factors in establishing and maintaining an effective program.

Interested third parties such as environmental, consumer, community, or other groups are continuing to hold businesses responsible for their environmental impacts and can demand improved environmental performance if they perceive environmental risks as unacceptable. ISO 14001:2015 requires that organizations externally communicate information relevant to the EMS, as established by the organization's communication processes and as required by its compliance obligations. A major strategic benefit of a carefully crafted external communications program is the development of cooperative relationships with stakeholders, which encourages their involvement in and fostering of an ongoing and productive dialogue regarding environmental activities and impacts.

3.10.1 Principles of External Environmental Risk Communication

Six basic principles of environmental risk communication should be applied to any external communication program developed as part of an ISO 14001 environmental management system:

Unfamiliar risks are less acceptable than familiar risks. Any external communications program should detail information on the pollution or waste streams, including where they come from, how they are integrated into the production process, what their environmental impacts are, and what the plans for continual improvement are. Comparisons with existing processes used safely by other businesses and industry standards also help to build credibility and better relationships.

Involuntary risks are less acceptable than voluntary risks. The communications program manager must acknowledge the stakeholders' involvement in environmental decisions to make them feel increased personal power and a part of the environmental decision-making process.

Undetectable risks are less acceptable than detectable risks. The communications manager must state clearly just what comprise proper and improper operations with regard to the business's environmental impacts. A description of the monitoring and measuring activities will greatly reduce tensions with unfriendly stakeholder groups.

Risks perceived as unfair are less acceptable than risks perceived as fair. The key to fairness is to respond appropriately to stakeholders regarding their perception of risks associated with the facility's operation. Response can take a wide variety of forms, including direct monetary contributions for industrial accidents, community support programs, and involvement with community and charitable organizations.

Dramatic and memorable risks are unacceptable. People tend to judge an incident as more likely to occur if they can easily imagine it or recall a similar instance. Oil spills in various parts of the world have been vividly portrayed on television. These events are experienced personally by a great many people who will fault businesses for negligent or voluntary conduct that results in environmental calamities. Companies need to first understand the possible consequences of their operations in order to clearly state the protective measures that the company is prepared to take to prevent the occurrence of an accident.

Stakeholders are less interested in risk estimation than in risk reduction, and they are not interested in either until their fears have been validated. Discussions about how to reduce risk tend to be more productive and demand a higher level of stakeholder involvement than those concerned with estimating the actual risk potential. Discussions should be focused on solutions rather than on theoretical possibilities, and on concrete action items rather than on placing blame. Careful discussion of the issues coupled with collaborative efforts to seek mutual goals is necessary to ensure the stakeholders do not have their fears validated.

The external communications strategy needs to consider the legitimate concerns of both environmental advocacy groups and consumers. These groups demand changes in environmental management practices that they perceive as posing unacceptable risks. Management must carefully plan and pursue communications initiatives targeted at such advocacy groups to reduce the potential development of costly, adversarial relationships.

3.10.2 External Communications Methods

Methods for communicating with external stakeholder groups include the following:

- Public meetings with communities and chambers of commerce
- Annual environmental reports
- Community advisory panels
- Fact sheets and facility awareness information packets
- Press releases and articles in newspapers, magazines, and trade journals

- Social media
- Media events
- Meetings with high-level public officials and regulators
- Customer service activities
- Facility tours
- Hotline numbers
- Trade association activities
- Websites or apps

Each of these methods is self-explanatory, with the exception of environmental reports.

3.10.3 Environmental Reports

Environmental reports are an excellent way to transmit a company's environmental message and performance record to key stakeholders, including the employees and the general public. These reports may also be used to demonstrate publicly the continual improvement of environmental performance of the company. A recent study of the 120 largest publicly traded companies in France during the period 2007–11 confirms that, while shareholders and consumers relate differently to a company's environmental reporting, both groups more favorably perceive firms that have better environmental performance, resulting in the firms' demonstrably increased sales growth, profit margins, and market value, particularly in industries positioned in close proximity to customers (Radhouane et al. 2018). One recent exemplar of environmental reporting is Graham Partners' 4th Annual Sustainability Report, which outlines their sustainability goals and accomplishments for 2018. Under the direction of Graham Partners' Director of Sustainability, the report provides a concise, clearly organized, and reader-friendly overview of relevant statistics, growth, initiatives, sustainable sourcing, case studies among their businesses, energy audits, internal policies and philosophies, and recognition of their internal impact award winners.

3.11 Environmental Communications Strategy

The development of the environmental communications strategy should follow the "plan-do-check-act" cycle of continual improvement. Each area must be defined and implemented to ensure comprehensive coverage of internal and external stakeholder environmental issues.

3.11.1 Plan

- Define the intent and purpose behind the program, including objectives and targets
- Define the scope, content, and type of the environmental communications (participatory or nonparticipatory)
- Identify the potential internal and external stakeholders and their interests and concerns
- Define a set of implementation activities necessary to match the communication of the business's environmental issues with each stakeholder's interests and agenda

3.11.2 Do

- Contact all stakeholders and obtain their involvement in and potential endorsement of the program
- Find out their specific agendas and interests
- Inform them about the business's environmental management activities
- Establish a format and mechanism for ongoing productive dialogues with stakeholders
- Establish a format and mechanism for community awareness and public and customer relations
- Establish involvement with ongoing community activities, such as civic groups, chambers of commerce, rotaries, nongovernmental organizations, and schools

3.11.3 Check

- Conduct periodic internal and external stakeholder surveys, community and customer focus groups, and public meetings
- Establish hotlines for environmental issues and monitor stakeholder data and issues previously discussed
- Monitor media coverage of environmental issues related to the business
- Evaluate customer response to environmental strategies
- Review public statements or reports by interested parties

3.11.4 Act

- Conduct reviews of the external communications program as part of the management review process
- Implement changes as necessary for emerging issues or areas of concern and to ensure continual improvement

Well-developed information and communications strategies can play an important role in environmental management systems. Equally important is the ability of companies to eliminate liability risk altogether by using effective risk transfer strategies. The following chapters discuss how this is accomplished.

References

Armao, J.J., and B.J. Griffith. "The SEC's Increasing Emphasis on Disclosing Environmental Liabilities." *Natural Resources and Environment* 11 (Spring 1997).

Graham Partners. *2018 Sustainability Report*, April 2019. https://www.grahampartners. net/media-center/.

ISO. "ISO 14001 Environmental Management Systems—Revision." https://www.iso. org/iso-14001-revision.html.

Kuhnhein, C., and K. Burke. "SEC Clarifies Environmental Disclosures: Climate Change." *Law Bulletin, Taft*, 2010. https://www.taftlaw.com/news-events/law-bulletins/sec-clarifies-environmental-disclosures-climate-change.

Radhouane, I., M. Nekhili, H. Nagati, and G. Paché. "The Impact of Corporate Environmental Reporting on Customer-Related Performance and Market Value." *Management Decision* 56, no. 7 (2018): 1630–59. doi:10.1108/MD-03-2017-0272.

4

Strategic Information for Risk Management Systems

Robert A. Woellner

Strategic information management is the integration of the complex array of environmental data existing within most businesses to support management decision making. Management needs accurate and complete environmental information to facilitate regulatory reporting and environmental communications and to increase the effectiveness of decisions regarding environmental performance. With information integration comes an increased awareness of environmental management activities—leading to the identification of opportunities to eliminate redundant or polluting operations, to minimize energy and natural resource usage, and to address high-risk issues. Environmental risks are identified in this process and managed on the basis of credible data rather than speculation.

4.1 Evaluation and Selection of Information Technology

Automated information technology supports an environmental management system and makes it possible to bring together the types of information necessary for effective environmental risk control. Even though the ISO 14001 standard does not specifically require information technology, it is difficult to imagine a business addressing its environmental activities without some form of automation. What the standard does require is a clearly stated and comprehensive description of how the business manages its environmental performance. Information technology makes such a description possible and assists in managing the risks associated with the daily activities of all employees. Once environmental information is collected and analyzed using a database, steps can then be taken to reduce risk, as demonstrated in later chapters -of this book.

Before considering which specific information technology to use, companies should first evaluate the following overall performance criteria required by an environmental management system. The technology must:

- Provide complete information about the environmental performance of the business
- Meet all current regulatory and organizational reporting needs
- Establish links among the different environmental management system elements
- Provide a wide variety of personnel with access to functional information about the elements of the environmental management system
- Support the decision requirements of personnel responsible for environmental management, as defined and documented in the environmental management system
- Support the transaction needs of each environmental management system element

Management must also consider three basic technological strategies to support its environmental management system:

- Modification and enhancement of existing systems
- Purchase and installation of a commercial system
- Development and implementation of a data warehouse–based system

Many businesses already use some form of environmental management information system. These systems, however, tend to be fragmented and optimized for particular functions, such as for the federal Resource Conservation Recovery Act (RCRA), hazardous waste management, purchasing, or regulatory reporting. Numerous technologies also purport to provide integrated information support for an environmental management system but can often cause more problems than they resolve. Too much information in a decentralized context, or seemingly inconsistent data packages, can confuse decision makers and cause conflicts and disruptions in business operations.

Should management build an entirely new system based on the new environmental management system requirements? While there is no simple answer, the existing information systems should be evaluated against the six performance criteria listed above. An environmental management system by its very nature must be customized to the activities and processes of the business.

Modifying or enhancing existing systems may be appropriate for recently installed or upgraded systems, but the integration requirements of the environmental management system may exceed the capacity of those systems or require reduced performance of individual functions. Businesses need to keep in mind that commercially available environmental management system packages need to be connected to the existing systems. A canned environmental management system package is not appropriate for mid- to large-sized businesses with complex environmental regulatory obligations.

Many large businesses implement enterprise-wide solutions. The business creates a series of transaction-based modules that feed data into a central database or a series of databases. The management can then access the databases by using a high-level query tool with the capacity to deal with complex or technically challenging issues. This type of solution can support records management, communications, and data trending and analysis for continual improvement of environmental performance.

4.2 Process Reengineering

Any information technology development project should involve some form of process reengineering if it is to provide value to the business. An ISO 14001 environmental management system must integrate the environmental function into all relevant business processes so that they can be managed within the context of the strategic positioning of the firm.

During the initial analysis, redundant or inefficient activities in the process should be reduced, often with the help of an information technology systems analyst. It is much easier to support the operation of a rationalized process than to attempt to retrofit a solution onto a flawed procedure. Thus, management can assure itself that it has measured its processes against environmental criteria and has built-in information technology solutions to integrate those processes on an organization-wide basis.

Other significant process management issues must be considered. Increased availability of information will change the way decisions are made about environmental performance and can have tremendous collateral consequences as job functions change and as employees seek new ways to exercise decision-making powers. Those who do not innovate with increased information and who continue to use old decision processes may not stand up to the scrutiny of the systems analysis efforts. Excuses like "that's the way we've always done it" will become unacceptable.

The new set of decision processes will be made by employees who have access to more information about how the business works than ever before. These decision makers may have to work with strangers and make decisions that may seriously affect other departments and functions. If the business is not already integrated and cross-functional, the information technology solution supporting the environmental management system will force these issues and this may have tremendous cultural and political implications for overall organizational performance. Managers who have built empires around the control of proprietary information will find them crumbling as they are exposed to the web of highly integrated information.

A central issue for larger businesses is integrating data across corporate levels and within each business unit. This will reveal duplication of effort and redundant operations. Data integration requires, however, semantic consistency for all business units. This means standardizing the way data are defined, gathered, processed, and reported in order to establish a useful basis of comparison between operations or sites. Semantic consistency is also critical during the data compilation and reporting stages to ensure that meaningful summaries are prepared for senior management and to ensure that errors and dysfunctions are identified and corrected quickly.

Another process management issue is the need for training and awareness regarding how the business manages its environmental impacts. In order to function within an integrated environmental management system, the employees need to be personally aware of their individual impacts on the environment. This includes how they travel to and from the workplace; their personal energy consumption; and how they integrate recycling, energy minimization, and other waste reduction technologies into their daily business lives. Information obtained concerning their environmental performance, both individually and collectively, can improve the decisions they make and can impact the overall functioning of the business.

Employees also need to structure decisions to achieve maximum environmental performance. Thus, training takes on several new dimensions beyond specific job functions, the most important of which is the management of innovations and change. Employees need to comprehend how management has chosen to structure the environmental management practices of the business, as well as their roles and responsibilities for continual improvement. Employees also need to know how their jobs are affected by environmental practices, what they are expected to do, and the reasonable expectations for organizational environmental performance. Employees' job performances should be analyzed on a yearly basis, and job evaluation criteria should include how they have been able to make progress in reducing their environmental impacts at the workplace. Companies can consider bonuses and other incentives for employees who substantially reduce their individual impacts. All of this information can be made a part of a central data system.

4.3 Consideration of the Elements of the Environmental Management System

Regardless of the information technology system selected, each element of the ISO 14001 standard also has specific requirements for the creation, dissemination, and retention of environmental information. The significance of each element and its attendant performance characteristics are detailed below.

4.3.1 Environmental Policy

An environmental policy is an excellent vehicle to set forth, for internal and external review, a company's capabilities to collect information on its own environmental impacts. In creating a sound policy, management teams, in conjunction with employees, need to develop an information technology strategy that provides for the dissemination of policy information to employees at all levels, the public, and other stakeholders. The policy can identify various communications devices like social media, e-mail, or file-sharing communications functions to assist this effort.

4.3.2 Environmental Planning

The information technology strategy should support the business's environmental aspects and provide a means of describing processes and their relationships. Especially important is the inclusion of a life-cycle modeling capability to perform sensitivity analyses to determine where and how environmental impacts can be minimized or eliminated.

The information technology strategy can support the tracking and evaluation of legal requirements, including a method to evaluate the demands of changing regulations. The strategy also needs to incorporate organizational requirements that impact environmental performance, such as community commitments, agreements, and consent decree actions.

The automated information technology can greatly assist management with the control of objectives, targets, and programs. The strategy should be able to identify:

- Current goals, objectives, standards, and trends of industry
- Progress against those goals, objectives, and industry averages (benchmarking)
- Priority activities and their relationships to other areas of organizational activity (marketing, sales, finance, or operations)

4.3.3 Environmental Operations

The information technology strategy should provide an interface to the environmental training function and report on the current status of each employee's environmental training and job performance. Staffing requirements, scheduling, and selection need to be made available, as well as a means for evaluating skills and placing personnel in appropriate positions.

4.3.4 Emergency Planning and Response

The information technology strategy should support the documentation requirements of the environmental management system, including regulatory reporting, recording of operating conditions and environmental exposures to employees, and both overall and process-specific emissions levels. A key consideration is the integration of all documentation activities related to environmental performance, including such areas as market research and customer requirements, product and service design, subcontractor selection and evaluation, manufacturing process development and control, and product use and disposal operations.

The information technology strategy should support emergency preparedness and response activities, including sensors to detect excessive toxic releases. The management of emergency response activities can be assisted by site mapping, personnel identification, hazard determination, and external communication functions.

4.3.5 Monitoring and Measuring

Integration of data obtained through automated data collection by process-monitoring equipment, such as sensors and detectors, is crucial to obtain a clear picture of environmental impacts such as emissions. The information technology strategy should also include functions for tracking and trending results of compliance and environmental management system audits, as well as the corrective actions resulting from these audits.

4.3.6 Management Reviews

The information technology solution should assist management by gathering and reporting all the data generated by the environmental management system into some form of executive information system for review. The necessary capability will bring together diverse sets of data to support strategic decisions regarding environmental performance and the management of attendant risks.

The information technology strategy is an important component of the environmental management system. Information that is mishandled or miscommunicated can be devastating to system operations. Companies need to design effective risk communications components to complement the information technology system.

5

Voluntary Programs and Industry-Led Initiatives to Reduce Environmental Risk

Robert A. Woellner

Businesses have various mechanisms to choose from to identify, assess, and manage environmental risk. Properly implemented environmental management systems can effectively accomplish all three of these business goals. Insurance is a critical component of any business operation. This chapter discusses how companies can manage environmental risk by transferring all or part of their environmental liabilities to third parties. Choosing the correct risk transfer strategy is a crucial business decision. Recently, many new options have emerged to transfer environmental risks to third parties.

Risk transfer can be closely related to a business's risk financing strategy. Each business must determine its tolerance for the self-assumption of risk and identify appropriate methods to distribute or transfer the remainder of the risk. A business choosing not to assume all of its own risks can transfer its current or potential environmental risks through a variety of mechanisms, including insurance, contractual risk transfer, and property transfer. These risk transfer mechanisms are not mutually exclusive, and the most cost-effective strategy is usually a combination of the various alternatives.

5.1 Environmental Insurance

Insurance is by far the most common risk transfer mechanism. Billions of dollars are paid in premiums each year to transfer exposures to the insurance community. Businesses began purchasing general insurance in the late 19th century, with each insurer issuing its own policy language. In the 1930s, the insurance industry began to standardize policy language, and in 1941, the first comprehensive general liability (CGL) policies were introduced (Bailey and Gulledge 1997). For many years these policies were purchased by businesses and resulted in insurers being held liable for environmental claims arising from waste disposal during this time period. In the early 1970s after Congress began enacting comprehensive environmental laws, the CGL policies were revised to exclude property damage from pollution or contamination that was not sudden and accidental.

The pollution exclusion provision contained in most general liability policies issued between 1973 and 1986 included the following language:

> This insurance does not apply to: bodily injury or property damage arising out of the discharge, dispersal, release or escape of smoke, vapor, soot, fumes, acids, alkalis, toxic chemicals, liquids or gases, waste material or other irritants, contaminants ... , but this exclusion does not apply if such discharge, dispersal, release or escape is sudden and accidental.

With the introduction of the absolute pollution exclusion and the total pollution exclusion clauses in commercial general liability policies, it is now necessary to add pollution coverage endorsements to these forms or to purchase separate environmental liability insurance if environmental risk is to be insured. Environmental impairment liability insurance came into existence in 1977 as separate insurance coverage; however, the market for environmental insurance remained relatively restricted until the late 1980s. With the opportunity to underwrite environmental risks, many insurers in the standard commercial property and casualty insurance market did not routinely add pollution coverage to general liability insurance policies, leaving the policyholder liable for environmental damages and restoration. If environmental coverage was provided by the general insurance underwriter, it was commonly time element or sudden release coverage that did not include claims for damage to soil or groundwater resources.

Since 1986 the market for specific types of environmental insurance coverages has expanded rapidly. Currently, more than 25 varieties of environmental insurance are available, and more than 100 different environmental insurance policies are offered by pollution underwriters. Insurance products are available to cap remediation costs, to protect buyers and sellers from environmental liabilities resulting from property transfers, to protect owners from known and unknown environmental impairments, and to control insurance costs of environmental service providers. New forms cover claims arising from sudden and accidental, as well as gradual, releases. The following specific environmental insurance products are currently available:

- **Environmental Asset Liability (EAL):** Coverage to protect property owners from the liabilities associated with known and unknown environmental conditions preexisting at a site.
- **Environmental Impairments Liability (EIL):** Coverage to protect property owners from the liabilities associated with ongoing operations.

- **Remediation Stop Loss:** Coverage to cap remediation costs for known environmental conditions and limit risks to the owner. This cost cap coverage is designed to insure remediation costs that exceed the projected or anticipated costs in the execution of a remedial action plan at a specific location. This coverage is particularly useful in facilitating the sale of contaminated property.
- **Owner Control Insurance Program (OCIP):** Coverage for construction or remediation activities. This coverage is held by the owner rather than the contractors in order to control costs.

Other types of coverage include:

- Contractors' pollution liability coverages
- Engineers' and consultants' professional errors and omissions liability coverages
- Asbestos and lead paint abatement contractors' insurance coverages
- Combined forms for professional and contracting risks

In addition to standard insurance coverages, environmental insurance policies are currently available to cover an extensive list of possible losses, including those arising from:

- Emergency response actions
- Government actions
- Site assessment related to a pollution incident
- Remedial cleanup costs, including leaking underground storage tanks, PCB spills, and spills from stored hazardous materials
- Business interruption costs
- Monitoring costs
- Third-party litigation costs
- Remediation closure costs
- Change in state or federal laws
- Real estate stigma (diminution in value, guarantee of fair market value, etc.)
- Interim ownership securitization for the chain of title

The environmental insurance industry is in an exceptional state of transition. Any business that has not addressed its environmental insurance options within the past five years may be paying too much for too little coverage. Although there are off-the-shelf insurance products available, the best way to insure complex environmental risk is through a customized menu of insurance products designed to minimize a business's liabilities while not inflating premiums with unneeded coverages.

The implementation of an environmental management system such as ISO 14001 should favorably impact the cost and availability of environmental liability insurance products. Businesses that demonstrate environmental proactivity through conformance with ISO 14001 should receive favorable underwriting consideration and may qualify for a decrease in premium levels.

5.1.1 Contractual Risk Transfer

A contract is an agreement between two or more parties supported by adequate consideration which creates a duty. The intent of a risk transfer contract is to have the risk reside with the party who has responsibility for action. However, the risk transfer mechanisms are not always put into writing prior to the loss, with the parties acting responsibly to resolve the loss according to the agreed-upon loss-sharing arrangement. Any lack of documentation or ambiguity in language will provide an opening for litigation. Often a judge or a jury allocates the loss. Unfortunately, when the effectiveness of a contractual risk transfer arrangement is determined by litigation, it can result in unpredictable and expensive outcomes.

Properly written, contractual risk transfer allocates the loss potential of an activity or practice in a fair and equitable manner. The general theory is that "I will be responsible for my employees and property, and you will be responsible for your employees and property." Damage to third parties should be the responsibility of the person in control of the activity or operation. These simple concepts become complex when they are considered in the contexts of unequal bargaining power; federal, state, and local environmental laws and regulations; and changing market conditions. Contractual risk transfer techniques include hold harmless agreements, waivers of subrogation, exculpatory agreements, indemnity agreements, and contractual definition and division of liability agreements as outlined below.

Under a hold harmless agreement, a contracting party will assume the legal obligations of the other party from third-party actions. The hold harmless agreement becomes effective when one party (the indemnitor) agrees to hold the other party (the indemnitee) harmless from tort liability under third-party actions associated with the hazards of the venture. By virtue of the agreement, the indemnitor assumes the liability of the indemnitee. In a broad form indemnity, the indemnitor will be responsible for any and all liabilities, regardless of which party was at fault. In an intermediate form indemnity, the indemnitor assumes all liabilities with the exception of those due to the indemnitee's sole negligence. Under the limited form indemnity, the indemnitor owes the indemnitee for that portion of the claim related to his own liability. Although the limited form indemnity may make the most logical sense, it often leads to lawsuits where the courts apportion the claim since the percent or cause of liability will often be in dispute.

Waivers of subrogation are agreements between two parties that state that one or both parties' respective insurance companies have no right of recovery against the other party. Essentially, the insurance companies are made a part of the contractual agreement between the two parties. In this way, the parties can be certain that their insurance companies will not sue the responsible or negligent party, thereby circumventing the hold harmless agreement found in the contract. Generally speaking, whenever there is a hold harmless agreement in a contract, it should also be accompanied by a waiver of subrogation provision.

In an exculpatory agreement, one party accepts the entire blame for the actions of the other party and handles the entire loss or claim just as if the other party were nonexistent. Exculpatory agreements are common in equipment leases, mineral leases, land leases, right-of-way agreements, and water use agreements.

An indemnity agreement protects against, and pays for, possible damage, legal suit, or bodily injury. Indemnities commonly involve the owner of personal property and the business that has temporary possession of the property. The holder of the property agrees to pay the property owner for any physical damage losses stemming from the acts or omissions of the holder of the property.

The contractual definition and division of the liability risk transfer technique is an agreement specifying who will be financially responsible for certain losses. This type of agreement identifies potential losses and who will bear the financial responsibility for these losses.

5.1.2 Property Transfer

Many owners of contaminated property decide to divest themselves of environmental liabilities by selling their properties. These owners must realize that in the U.S. environmental laws impose joint and several liability for cleanup costs on all parties in the chain of title when there is a release. Therefore, selling a property may not reduce liabilities and may even increase exposure due to the likelihood of new discoveries. Sellers therefore desire to transfer properties to parties that have the resources not only to remediate the property but also to protect the seller from possible future costs related to environmental liabilities of the property. For similar reasons, potential purchasers are extremely cautious about taking title to contaminated property.

When selling contaminated property, buyers and sellers often have difficulty in establishing the sales price, since there is commonly a wide discrepancy between the low and high estimates of cleanup costs. Whereas potential buyers tend to discount the sales price by the maximum potential remediation cost, the seller favors the low-cost estimate. Many real estate deals fail because of this lack of agreement. In such instances, remediation stop loss or cost cap insurance can be useful in facilitating the sale of the contaminated property. Remediation stop loss policies agree to pay, on behalf of the named insured, the expenses (in excess of the self-insured retention) that the insured incurs in completing an approved remedial activity at a specified location. Although the insurance has a price, using cost cap insurance can provide a remediation cost number that is guaranteed and therefore useful to both the buyer and the seller.

The existence of risk management systems and new environmental insurance coverages has made possible a new industry of companies that buy contaminated real estate, known as brownfields, with an agreement to hold the seller harmless from all future environmental liability. The risk assessment, control, and management expertise of these risk assumption companies allows them to approach municipalities and regulators with an environmental action plan that focuses on reduction of environmental exposures. A municipality may be able to indemnify the new owner or occupant for historical contamination after remediation is completed, and insurance can supplement the indemnification, protecting the municipality against future claims associated with historical contamination. The lender providing remediation or purchase funding also can be protected against third-party claims (Bailey and Gulledge 1997). This demonstrated ability to minimize exposure rationalizes the process and allows government officials to exercise whatever latitude they may have to expedite real estate transactions. Buyers of contaminated properties have multiple options to protect their interests, including remediation cap insurance covering the buyer for cleanup costs exceeding the site estimate; historical and ongoing coverage protecting the buyer from remediation costs associated with unknown and undiscovered contamination; and insurance for known releases and future site operations (Bailey and Gulledge 1997). The next section discusses brownfields in greater detail to show how environmental management systems can be used to improve the decision making related to management, sale, and redevelopment of these contaminated properties.

Brownfields present complex questions of social, political, and economic policy for governments and businesses throughout the world. The expansion of insurance coverage for past and future environmental contamination, coupled with the change in the U.S. regulatory climate, has caused new policies to develop which are designed to provide incentives to businesses to redevelop contaminated properties. In this section we take a closer look at brownfields, the opportunities they present, and methods to effectively manage risk along the way.

5.2 The Brownfields Problem

Brownfields have been defined in federal legislation as "parcels of land that contain or contained abandoned or under-used commercial or industrial facilities, the expansion or redevelopment of which is complicated by the presence or potential presence of hazardous substances, pollutants, or contaminants" (Community Revitalization and Brownfield Cleanup Act of 1997). Developers' past avoidance of brownfields because of uncertainties in the amount of cleanup and development costs have contributed to construction on undeveloped greenfields sites, which create urban sprawl and infrastructure problems and reduce the amount of open spaces. Brownfields exist in every country in the world and pose remarkably similar problems.

Some countries, like the U.S., are attempting to solve these problems by remediating existing brownfields and, at the same time, seeking to eliminate the causes of contamination in an effort to prevent or at least reduce the number of future brownfields. As mentioned in previous chapters, the U.S. has been using command-and-control environmental laws and regulations for many years to force companies to clean up past contamination whether they were responsible for it or not. Many businesses are reluctant to expend resources on old industrial sites and instead decide to move their companies to new locations, usually uncontaminated greenfields, or to simply shut down operations.

One exception is Howard M. Kilguss, the owner of Excell Manufacturing (a $40 million jewelry manufacturing company located in Providence, Rhode Island). Mr. Kilguss decided not to rebuild in a greenfields location outside the city, but instead relocated two of his jewelry factories to an industrial site within the city. Kilguss was concerned about the environmental implications of the move, which was made necessary because of the need to consolidate both factories. He was also worried about the impact on many of his employees who took public transportation to get to their workplaces in Providence. In relocating both factories to one industrialized site in the city, he measured the environmental, social, and political impacts of the move and decided that it was more important to reduce those adverse impacts than increase short-term profits. Kilguss went further and had his employees actually participate in the planning of the new, larger facility in an attempt to have all the employees buy into the environmental considerations that underpinned the project. Employee support made the transition work better in the long term. Not all company owners have the same concerns for their employees or for the environment.

According to government studies, there are more than 450,000 brownfields sites in the U.S. (EPA 2016). In 1995, the U.S. EPA launched its Brownfields Program to provide small amounts of money to local governments to assist with hundreds of two-year pilot projects. In 1996, President Bill Clinton announced a $2 billion plan to clean up brownfields. He stated that this project was "the most important thing I am working on with the mayors of America today." The government's commitment to cleaning up brownfields has continued to evolve since then. In 1997, the Clinton Administration announced the creation of the Brownfields Partnership of 15 federal agencies, which was expected to leverage from $5 to $28 billion in private investment, support up to 196,000 jobs, and protect up to 34,000 acres of undeveloped greenfields on the outskirts of cities throughout the country. In 2002, the Small Business Liability Relief and Brownfields Revitalization Act expanded the EPA's assistance in both public and private sectors; amended and clarified the Comprehensive Environmental Response, Compensation, and Liability Act (CERCLA or Superfund) by allocating funds toward assessment and cleanup and by defining liability protections; and also provided funds for state and tribal programs. Most recently, the 2018 Brownfields Utilization, Investment and Local Development (BUILD) Act "reauthorized EPA's Brownfields Program, and authorized changes that affect brownfield grants, ownership and liability provisions, and State & Tribal Response Programs" (EPA 2018b). In the 2018 fiscal year, the Brownfields Program exceeded all of its fiscal year targets, with 1,919 properties assessed; 11,197 jobs leveraged; $2.201 billion leveraged; and 7,986 acres made ready for anticipated reuse. Cumulatively over the life of the program (through October 1, 2018), these figures tally up to 28,968 properties assessed; 141,332 jobs leveraged; $26.807 billion leveraged; and 77,038 acres made ready for anticipated reuse (EPA 2018a).

There are some compelling economic reasons, aside from health and safety issues, why government is seeking to clean up these sites. Many of these unremediated brownfields are in urban industrialized areas. There are obvious economic benefits to redeveloping these sites. For example, the redevelopment of brownfields in cities will create jobs and place them closer to dense population centers, thereby reducing urban unemployment and unnecessary transportation problems. "The mismatch between residential locations and the location of jobs is a problem for some workers in America because, unlike the system in Europe, public transportation is weak and expensive" (Wilson 1996). Inner city redevelopment will increase municipal tax bases and lessen future suburban sprawl, according to the U.S. Conference of Mayors, which recently completed a survey in which 84% of the respondent cities said that they had successfully redeveloped a brownfield site. In addition, 106 cities reported the creation of 187,000 jobs due to redevelopment of brownfield properties, 116,000 of those jobs being permanent. These new developments have naturally led to increased local, state, and federal tax revenues, as the survey revealed that 62 cities reported that actual tax revenues from redeveloped brownfields sites totaled more than $408 million. Moreover, these cities estimated their potential revenues as ranging from $1.3 billion to $3.8 billion (Bollwage 2017).

For years industry has been relocating to undeveloped rural areas, at times creating new polluted areas and requiring expensive extensions of infrastructure and municipal services to previously undeveloped land. In Chicago, for example, between 1970 and 1990 the regional population increased only 4% while the availability of urban land area expanded by 46%. The development of open space and farmland can have a negative impact on farming, natural habitat, air quality, energy consumption, and congestion. During that same period of time, total vehicle miles traveled in Chicago almost doubled because of sprawling land-use patterns. A costly public transportation infrastructure was required to accommodate the increased dependence on the use of automobiles. This particular scenario resulted in the Northeast Illinois area being in severe noncompliance with federal air quality standards. Six of the seven chief air pollutants come from automobiles, according to the EPA (Anonymous, 1997).

The air quality problems due to excessive automobile usage in the Chicago and Northern Illinois areas are consistent with national trends. The Federal Highway Administration reported that in 1970, U.S. drivers covered approximately 1,120 billion vehicle miles; the annual total had risen to 2,829 billion vehicle miles by 2002 and to 3,213 billion vehicle miles by 2017 (FHA 2019). By 2016, 128 million Americans (85.4% of workers) commuted to work by car, with 119.5 million of them (76%) being the sole occupant of the vehicle. Although the number of car commuters in 2016 was 1.6% lower than in 2006, it represents an increase of 8.5 million people as compared to the 2006 figures (Loesch 2017; Tomer 2017). By 2016, the US Energy Information Administration recognized that transportation had become "the economy's number one polluting sector" (Tomer 2017).

Federal, state, and local governments need to focus on transportation problems in conjunction with finding ways to redevelop brownfields so that jobs can be created closer to the labor market. New coalitions are now being formed to develop programs and undertake other initiatives to find solutions to these environmental-labor problems. While current politically popular initiatives mostly focus on how to clean up existing brownfields, very little attention has been paid to developing systems to manage and prioritize the redevelopment of brownfields, to eliminate the factors that could lead to future brownfields, and to create opportunities for continuous improvement of the redeveloped land. A key component of any brownfields redevelopment management plan is making certain that the businesses stay in their recycled homes so that properties have sufficient labor resources and unnecessary travel is reduced.

5.3 Brownfields Remediation

In February 1995 the EPA began its Brownfields Initiative, by focusing not on labor and transportation impacts, but on the stigma attached to these properties. The EPA deleted approximately 28,000 properties from its Comprehensive Environmental Response, Compensation, and Liability Information System (CERCLIS) database, based on its determination that these sites had a low likelihood of requiring further remedial action. The EPA determined that there was "no further remedial action planned" (NFRAP) at these sites under federal environmental law and acknowledged that maintaining the sites on the federal CERCLIS database had made it more difficult for lending institutions and real estate developers to redevelop these properties. The EPA decided "to counteract the market perception that, by being listed on CERCLIS, a site must have a significant hazardous waste problem and therefore should not be considered as a potential redevelopment site" (Abelson and McCaffrey 1996).

Between 1995 and 1997, the EPA made grants to 113 communities across the nation to clean up the brownfields sites (Gore 1997). The first project, in Cleveland, Ohio, the restoration of part of an old industrialized property, created 200 new jobs and generated more than $1 million in new payroll taxes. In 2010, the EPA began awarding Brownfields Area-Wide Planning pilot grants to empower regional and local governments, tribes, and non-profit organizations to conduct brownfields planning projects in areas struggling with one or more brownfield sites. Besides area-wide planning grants, the EPA also offers (either through direct funding or in collaboration with other programs and agencies) the following resources: assessment grants; revolving loan fund (RLF) grants; cleanup grants; multipurpose (MP) grants; environmental workforce development and job training (EWDJT) grants; technical assistance, training, and research grants; and state and tribal response program grants. Investors, businesses, and developers should take advantage of these opportunities as many have already recognized the benefits of redeveloping these sites.

National policy initiatives to encourage brownfields development have been emerging. They include prospective purchaser agreements (PPAs). A PPA is an agreement wherein the EPA agrees not to sue (covenant not to sue) a prospective purchaser of a contaminated property if that individual or business organization agrees to clean up the site and if the PPA results in a substantial direct benefit to the EPA. The EPA defines such ancillary benefits as the creation or retention of jobs, productive use of abandoned property, or revitalization of blighted areas.

As detailed in the Colorado Department of Public Health and Environment (CDPHE) Brownfields and Land Revitalization website, the EPA's Brownfields Program empowers states, communities, and other stakeholders to work together to prevent, assess, safely clean up, and sustainably reuse brownfields. On January 11, 2002, President George W. Bush signed into law the Small Business Liability Relief and Brownfields Revitalization Act. Under the Brownfields Law, EPA provides financial assistance to eligible applicants through four competitive grant programs: assessment grants, revolving loan fund grants, cleanup grants, and job training grants. Additionally, funding support is provided to state and tribal response programs through separate mechanisms.

These federal programs and policy initiatives have been supplemented by administrative actions on the state and local levels. Some examples shown below will demonstrate a commitment by government at all levels to solve the brownfields problem.

5.3.1 State and Local Actions

A memorandum of agreement between the CDPHE and the U.S. EPA Region VIII states that the EPA and the CDPHE believe that the proper revitalization of contaminated or potentially contaminated industrial and commercial property (brownfields) will provide a significant benefit to both the environment and the economy of the local communities. To the extent possible, the EPA and the CDPHE seek to facilitate the productive reuse of brownfield properties by working with the private sector to eliminate impediments to financing, transfer, and redevelopment. The EPA and the CDPHE seek to encourage participation in the Colorado Voluntary Cleanup Program by allowing qualified owners of contaminated property to voluntarily propose cleanup actions and seek no further action determinations for eligible sites.

Massachusetts is another good example of innovative state brownfields programming. In 1994 that state launched a pilot initiative in which government officials worked with representatives from business and the private sector to encourage the redevelopment of contaminated sites in designated economic target areas (Abelson and McCaffrey 1996). As of 1996, Massachusetts had an estimated 7,000 brownfields sites. The intent of the initiative was to reduce the number of brownfields and to develop state policy to recycle these sites into productive properties. As of May 2017, the number of brownfields sites in Massachusetts had decreased to 1,012. Similar voluntary cleanup programs have emerged in many other states, including California, Illinois, Michigan, Minnesota, Missouri, Ohio, Pennsylvania, Colorado, Texas, and Wisconsin. In Massachusetts the applicant for the project must be a prospective purchaser or a tenant (not a "responsible party" with potential liability for the cleanup of the site) who is purchasing or leasing property in a disadvantaged economic target area. Massachusetts agreed not to sue the prospective purchaser or tenant if the property is cleaned up in accordance with Massachusetts law (the Massachusetts Contingency Plan or MCP) which addresses risk-based cleanup standards. The process has no specific state oversight. Licensed Site Professionals (LSPs) from environmental consulting firms are retained to move the sites through the MCP process. This new program won an award in 1995 from the Council of State Governments and is now being adopted by a number of other states, including Connecticut, Illinois, and Ohio.

Cities like Chicago that have huge numbers of brownfields are engaged in local efforts to remediate these sites and to revitalize significantly deteriorating urban areas. In November 1993 the Chicago Department of Environment, Planning, and Development; the Law Department; the Department of Buildings; and the Mayor's Office formed an interdepartmental working group on brownfields (Brownfields Forum 1995). The working group decided:

- To devise more responsive environmental and economic development policies

- To create a brownfields pilot program to clean up and redevelop demonstration sites in distressed neighborhoods
- To develop economic models that account more accurately for environmental and social cost benefits of development decisions

The working group found that the demand for industrial space in Chicago greatly exceeds its supply and that most of the hundreds of abandoned industrial properties within the metropolitan area are not competitive with greenfields sites in the suburbs and beyond. The working group further found that the fear of environmental costs and liabilities produced a stagnant marketplace where brownfields lay dormant for years. The working group chose five pilot projects to determine what steps would be necessary to recycle these properties for reuse. Of the five selected properties, one was found to be clean, while another had only minor contamination. When these and the other pilot properties were returned to productive use, the working group determined that these properties would create jobs that are especially valuable to the distressed communities where brownfields are commonly located. "Brownfields redevelopment can produce a halo effect, attracting additional investment in local businesses, public infrastructure and employment training" (Brownfields Forum 1995).

The working group further found that communities needed to "close the loop" that links environmental remediation, redevelopment, and redevelopment activities to reap the full development of brownfields cleanups.

> If there is one over-arching theme to the Chicago experience, it is the need for a coordinated, comprehensive effort involving all key stakeholders. No one group can solve this problem alone. City, state and federal agencies have an integral role to play. So do banking, business, manufacturing, legal, insurance and real estate professionals; community industrial and economic development groups; trade associations, environmental and public interest groups; environmental justice representatives; organized labor; and community health organizations.
>
> **(Brownfields Forum 1995)**

Following an extensive study, the Brownfields Forum came to these conclusions:

- Brownfields redevelopment should foster healthy communities throughout the city and region.
- Public incentives for greenfields development should not outweigh incentives for recycling brownfields. Redevelopment of brownfields areas will reduce the need for new infrastructure in outlying areas, conserve environmentally sensitive areas, and otherwise save the costs of sprawl.

- Engaging the private sector and expanding market resources are critical to brownfields redevelopment.
- Effective strategies require strong partnerships among government, communities, and the private sector.
- Public brownfields expenditures should:
  - Address sites that would not be redeveloped without government participation
  - Redevelop disadvantaged areas
  - Focus on areas where reuse of brownfields is likely to catalyze additional redevelopment
  - Create and retain jobs
  - Maximize public benefit
- Redevelopment efforts should seek to attract environmentally sound industries to prevent the spread of brownfields and to foster sustainable communities.
- Brownfields initiatives should be viewed as one important component of a comprehensive strategy for revitalizing urban communities.
- Brownfields redevelopment should seek to leverage broader, integrated strategies for providing viable, long-term, and area-wide development.
- Environmental cleanup standards must be clarified to accommodate a full range of land-use options.
- A large-scale brownfields redevelopment program should be based on knowledge and experience gained through pilot efforts and tests of innovative approaches and tools.

The federal, state, and municipal experience shows that all three levels of government are seeking to provide substantial financial incentives that will result in the redevelopment of property and in the creation of new jobs. One common theme among all these programs is that government officials have endeavored to study the brownfields problem in urban areas to understand what needs to be done to clean them up.

5.4 ISO 14000 and Brownfields

The ISO 14000 standards may be used to facilitate this planning process, speed up the remediation of these sites, and manage the properties or recycle them for future use. ISO 14000 environmental management systems may assist these government entities and responsible businesses in addressing the brownfields problem on a massive scale and in using scarce resources to recycle these properties for productive use. Before money is wasted due to poor planning and excessive transactional costs, it would be useful to have a policy that would allow for cost-effective and permanent solutions to brownfields problems.

Businesses that own or are responsible for brownfields can benefit by taking stock of the number of these sites and by devising the methods and means to manage these properties effectively until they are remediated on a prioritized basis, using risk-based corrective action, and put back into productive use.

No business with brownfields sites should embark on a brownfields redevelopment program without planning a comprehensive and consistent approach to manage the multiple risks associated with the problem. A business seeking to understand its brownfields problem, or a business intending to redevelop another company's brownfields properties, can use a formal environmental management system to develop an environmental policy, objectives, and targets to effectively manage the risk involved.

An EMS-brownfields redevelopment program would have as its goals increased productivity of property; creation of jobs for nearby residents; reduction of costs, wastes, and energy usage; prevention of pollution and unnecessary further industrial development; and assurance of responsible future use of the property. The goals must be clearly articulated in a policy by owners and managers who are seeking to develop appropriate standards to revitalize these distressed properties. Partnerships and joint ventures can be established to foster swifter and more cost-effective solutions to brownfields problems. ISO 14001 provides the foundation for these important business-social-economic decisions to be made.

The objectives of the program are (1) to reduce risk to abutters, neighbors, and employees if there are ongoing operations, using risk-based corrective action standards, and (2) to improve these properties for future productive use. Risk-based corrective action simply means that the cleanup standard is dependent on the future use of the property. More stringent standards are used for parks and residential areas as opposed to paved-over factory or commercial sites. Extensive outlet shopping malls are being built on brownfields in Carson, California, and Elizabeth, New Jersey; these projects, and others on a larger scale, need careful planning. ISO 14001 can provide the necessary template for a major redevelopment project.

Developing a risk management system is an essential step in providing a framework for solving the brownfields problem. In a multiple-site project, managers would begin by inventorying and categorizing the properties on the basis of location, size, use (present and past), identity of owners and previous ownership, contamination in soil and groundwater, source of contamination (on-site and off-site), cost of remediation, and future uses. A brownfields database would be the centerpiece for a management program that would be developed and implemented to restore properties or sell them off to businesses or venture capitalists. For those properties that would be retained in the business portfolio, policy planners and senior management can use EMSs to guide their decision making to manage or transfer existing liabilities on the property. A liability and insurance inventory is commonly performed in this process. Customized environmental liability insurance policies, as discussed in the previous chapter, may need to be purchased if properties are to be maintained.

Brownfields joint ventures have been booming in popularity and success in the U.S. Numerous real estate service firms have joined with environmental remediation firms to propose real estate ventures. For each of these major joint ventures, an EMS can guide decision making and encourage the development of a cost-effective, risk-astute project.

A brownfields redevelopment program with an EMS component can also fulfill many business objectives. Such a program can:

- Prevent redeveloped sites from becoming eyesores again through current or future mismanagement
- Reassure current owners, developers, lenders, and investors that their liability protection will not be compromised by operations of new enterprises
- Offer a new way for brownfields entrepreneurs to achieve environmental compliance without excessive government involvement
- Convince government officials and the skeptical public that a trade-off of a lesser level of current cleanup for a potentially greater level of environmental consciousness in future site operations is worthwhile

An important component of any good environmental management system is continual improvement. There is a significant need not to repeat the mistakes of the past. The vast efforts of the federal, state, and local authorities should not be wasted by repeating the mistakes of 1960s-style urban renewal. Environmental issues have become an important part of community awareness and planning. Old approaches to urban development will not work and may become a barrier to successful revitalization of urban projects. Three principles are raised in this context by Larry Charles, a noted community activist in Hartford, Connecticut, who is in favor of brownfields redevelopment as long as the community (1) is responsible for the success of the project; (2) is allowed to control the project; and (3) is held accountable for the results of the project (Charles 1997). In short, urban planning cannot be successful without significant involvement of the community that is going to be both responsible for and the ultimate beneficiary of the project.

EMS-brownfields redevelopment and management projects must not only avoid land-use problems that have created industrial waste but also improve the future use of the properties. Energy efficient green buildings, open space, conservation easements, new urbanism, and other innovative planning techniques make important contributions to the quality of life in all communities, but particularly those that have been impacted the most by industrialization. Planning and development must occur with coordination and significant input from transportation experts to make projects pedestrian friendly and promote the use of alternative modes of transportation, such as buses, light rail, van pooling, and bicycles. Persons familiar with EMSs are uniquely qualified to provide sound management advice to the communities and property owners, who must work with a range of experts to create projects that will prove to be sustainable.

Carefully planned efforts will undoubtedly attract environmentally sound industries that will prevent the future spread of brownfields and encourage environmentally and economically sustainable communities. A formal EMS, used in conjunction with economic planning processes, will permit consistency and conformity with accepted standards to replace haphazard industrialization that occurred before people were aware of or cared about the long-term impacts of pollution. Properly utilized, EMSs can assist businesses to remediate brownfields and turn these sites into productive properties that will be continuously improved in the future.

References

Abelson, N., and M. McCaffrey. "Brownfields: Recent Massachusetts and Federal Developments." *Environment Reporter*, March 15 1996.

Anonymous. "Worth Considering." *Colorado Commons*, Spring 1997.

Bailey, K.D., and W. Gulledge. "Using Environmental Insurance to Reduce Environmental Liability." *Natural Resources and Environment*, Spring 1997.

Bollwage, J.C. "Building a 21st Century Infrastructure for America: Revitalizing American Communities Through the Brownfields Program." *The United States Conference of Mayors*, 2017. https://www.usmayors.org/2017/03/27/building-a-21st-century-infrastructure-for-america-revitalizing-american-communities-through-the-brownfields-program/.

Brownfields Forum. "Final Report and Action Plan." October 1995.

Charles, L. "Community Involvement with Environmental Cleanup and Economic Development." *Paper Presented at the 1997 Connecticut Industrial Site Recycling Conference*, March 14, 1997.

Colorado Department of Public Health and Environment. "Voluntary Cleanup and Redevelopment Program." 2019. https://www.colorado.gov/pacific/cdphe/voluntary-cleanup.

Environmental Protection Agency (EPA). "Brownfield Sites." 2016. https://archive.epa.gov/pesticides/region4/landrevitalization/web/html/brownfieldsites.html.

EPA. "Brownfields Program Accomplishments and Benefits." 2018a. https://www.epa.gov/brownfields/brownfields-program-accomplishments-and-benefits.

EPA. "Overview of EPA's Brownfields Program." 2018b. https://www.epa.gov/brown fields/overview-epas-brownfields-program.

Federal Highway Administration (FHA). "Highway Statistics 2017." 2018. https://www.fhwa.dot.gov/policyinformation/statistics/2017/.

FHA. "Travel Monitoring: Traffic Volume Trends." 2019. https://www.fhwa.dot.gov/policyinformation/travel_monitoring/tvt.cfm.

Gore, A. "Brownfields are Common Ground." *Brownfields News*, April 1997.

Loesch, Dyfed. "How Americans Commute to Work." *Statista*, October 5, 2017. https://www.statista.com/chart/11355/how-americans-commute-to-work/.

Tomer, A. "America's Commuting Choices: 5 Major Takeaways from 2016 Census Data." *Brookings*. October 3, 2017. https://www.brookings.edu/blog/the-avenue/2017/10/03/americans-commuting-choices-5-major-takeaways-from-2016-census-data/.

Wilson, W.J. *When Work Disappears: The World of the New Urban Poor.* New York: Alfred A. Knopf, 1996.

6

Environmental Risk Management Systems: An Introduction

Christopher L. Bell

6.1 How It Used to Be Done

It is remarkable that there was a time, not long ago, when even relatively sophisticated organizations had to be persuaded of the benefits of identifying and managing environmental risks and compliance in a disciplined and systematic manner. Managing environmental risk was frequently viewed as the realm of the environmental professional, who would manage the somewhat mysterious and complex "black box" of environmental issues in relative isolation from existing organizational management systems. A common approach was "management by exception," where senior management did not really want to hear about "environmental" unless there was a problem. In the traditional situation, environmental professionals were placed in a very difficult situation, attempting to direct individuals over whom they had no authority regarding actions over which they had no control, yet were either implicitly or explicitly viewed as being responsible for environmental matters.

Environmental issues were frequently viewed through the relatively narrow lens of facility and operational compliance with legal requirements, focused on basic issues such as air emissions, water discharges, or waste management. These were post-production "externalities" that the environmental professional was typically expected to successfully manage with as little interference as possible with the core operations of the organization. In the traditional "silo" model, broader issues such as managing resources more efficiently (e.g., water and energy), material conservation and substitution, product stewardship (design, performance, and end-of-life), and an organization's place in the larger environmental and social context were not within the purview of the environmental function (and often were not on anyone's radar screen).

Several factors have, over time, acted to persuade most organizations that environmental issues merit the same degree and type of management attention as any other core business issues. These include the financial and reputational costs of adverse environmental incidents; the ever-increasing breadth, complexity, and burden of legal requirements; pressure from a broad range of stakeholders (including shareholders, employees, customers, regulators, and public interest groups); and an awareness within organizations of the positive value of sustainability and social responsibility to long-term survival and success (e.g., the "triple bottom line" and the "circular economy") (Elkington and Zeitz, 2014; Webster, 2014). This growing realization contributed to the long-overdue application of well-established management principles to environmental matters.

6.2 Management of Organizations

Managing environmental risk is a subset of the larger topics of managing organizations, business, risk, and compliance generally. While a full treatment of these topics is outside the scope of this chapter, a brief overview of operational management systems is useful for several reasons. First, they have been the subject of extensive practical experience and academic study around the world and across industrial and service sectors, providing a valuable resource that may be beyond the experience of most environmental professionals. Second, most organizations have already long been implementing various forms of performance, risk, and compliance management tools outside of the environmental sphere. Understanding how an organization successfully manages its core business challenges is essential for the effective and sustainable management of environmental risk and the potential integration of that effort into the organization's everyday operations. Otherwise, environmental risk management may be perceived as being external to "normal" operations. This can also lead to an organization not taking full advantage of its existing experience in successfully managing complex challenges. On the other side of the coin, sophisticated business management or operational systems that do not explicitly take environmental issues into account may suffer from a lack of credibility and could send a message that environmental issues are not part of an organization's core values or concerns. Lastly, at a professional development level, environmental managers who do not understand or are not comfortable with business management practices are likely to face career barriers.

It is widely considered self-evident that a core function of an organization should be explicitly, formally, and transparently managed, whether it be finance, purchasing, production, human resources, intellectual property, information technology, inventory, quality, distribution, or marketing. Whether it was double-entry bookkeeping in the late 1400s, the "scientific management" of Taylorism in the late 1800s, or Drucker's "management by objective" in the 1950s, organizations and academics have devoted inordinate amounts of attention to fine-tuning the effective operation of organizations (Chandler, 1977; Kanigel, 2005). The central point is that applying formal management principles to complex issues is nothing new (Drucker, 1993, 2001).

Consulting firms and business schools continue this tradition, churning out "new" business management theories or operational management systems (OMS) on a regular basis, with "lean management," "Six Sigma," "operational excellence," and "operations integrity" systems among the prominent models, and with technology companies advancing concepts such as "agile" project management and Amazon's 14 "leadership principles" (including "customer obsession" and a "bias for action") (Adkins, 2019; Charan and Yang, 2019; Miller, 2014). ISO has also been hard at work on management systems, beginning with the publication of the ISO 9000 quality management series of standards in 1987, followed by ISO 14001 on environmental management in 1996. ISO is now pushing out management standards on everything from anti-corruption (ISO 37001:2016) to business continuity (ISO 22301:2012) to food safety (ISO 22000:2018) to the recently completed standard on occupational safety management systems (ISO 45001:2018). There has been such a proliferation of ISO management systems standards that ISO has created the so-called Annex SL template to encourage standards writers to use a consistent management systems framework and has published a "handbook" titled *The Integrated Use of Management System Standards* (ISO, Second Ed., 2018-11).

At the risk of offending proponents of one theory or another, most of the management models incorporate similar key themes aimed at having an engaged and accountable leadership and empowered workforce that relentlessly deconstruct, evaluate, and improve work processes (or "value streams") to obtain maximum performance, efficiency, and customer satisfaction, using an array of data-driven analytical tools. They are also typically characterized by some version of the "plan-do-check-act/improve/correct" model that was popularized by the quality pioneer W. Edwards Deming in the 1940s and 1950s (Deming, 2012). Some of these approaches have been associated with specific companies, such as "lean manufacturing" and Toyota, "Operations Integrity Management System" and ExxonMobil, "Six Sigma" and the General Electric Company, and, more recently, Amazon's well-publicized "leadership principles" (Ohno, 1988; Pande, 2000). Other approaches are associated with specific sectors, such as the "agile" management style that was largely initially developed in the information technology sector. The propagation of these management trends bears some resemblance to the "coaching trees" of successful sports teams, where the methods of an unusually successful team quickly spawn a narrative of systemic management superiority that should and can be adopted by others. The narrative typically changes with the vicissitudes of success, as the methods of successive winners are anointed as best practices. As a practical matter, out in the field, organizations frequently implement some mix of the theories being peddled by the proponents of various models, tailoring systems to their circumstances.

Environmental risk and compliance management has been a relative latecomer to this party. Given the investments necessary to achieve environmental compliance, the downside risks of noncompliance and environmental incidents, the opportunities offered (and costs avoided) by achieving superior environmental performance or offering environmentally beneficial products and services, as well as the broad stakeholder interest in sustainability, one would think that disciplined and formal management of environmental risk and compliance would long have been a "no brainer." It is not completely clear why it took so long for organizations to apply accepted and well-tested management principles to environmental issues. It is certainly not because organizations do not know how to successfully manage complex issues: organizations have been doing this for generations.

While environmental issues have their own challenges and characteristics, there is nothing uniquely complex or difficult about managing them. If an organization can successfully manage the other operational and market challenges it faces, it is certainly capable of addressing environmental risk. It may be because environmental issues were long viewed as "externalities" and not inherent characteristics of an organization's functions. This frequently led to environmental issues being managed in a "silo" by technical and legal environmental specialists outside of the mainstream management of the organization's everyday affairs, frequently operating under the radar unless and until something went wrong. Sometimes the technical and legal environmental specialists themselves resist opening the "black box" of environmental expertise to proven management techniques, perhaps out of a belief that the traditional approach offers job protection or perhaps that there was something unique about environmental matters. This view is inconsistent with how many organizations have managed professional development in other disciplines, where "high-potential" individuals might be rotated through a variety of operational and staff positions as they move up the organizational ladder.

There is also an ongoing tension between the desire to push the responsibility and accountability for managing environmental risk into the operations that create the risk and the concern that operations may not place sufficient priority on environmental issues. From this perspective, the environmental function should retain a certain "independence" from operations. This tension reflects a long-held suspicion among some stakeholders, including in industry, that the "natural state" of most organizations, particularly in private enterprise, is to not take environmental issues seriously enough. The more cynical may conclude that many organizations did not take environmental seriously as a risk management issue because they did not have to, and only did so when the cost of noncompliance and mismanagement to their bottom lines and reputations became too high to ignore.

Several decades of relatively comprehensive environmental regulation and related government enforcement and stakeholder attention, along with the general awareness of the holistic importance of environmental issues (sustainability, "triple bottom line," "circular economy," etc.), should have by now cured these deficits. Just one example of this more holistic understanding of the place of environmental issues is the Business Roundtable's revised "Statement on the Purpose of a Corporation," issued on August 19, 2019, that adopted a broader and more inclusive view of the role of corporations, including with respect to environmental, sustainability, and social responsibility issues. John Elkington, known in some circles as the "Godfather of Sustainability," has expressed a similar broad multi-stakeholder approach as "regenerative capitalism" (Elkington, 2020). The need for comprehensive and agile responses to the COVID-19 pandemic and related public health issues has only served to emphasize this point. There has also been a convergence between organizational management and environmental management, in terms of both personnel and management tools. It is increasingly common to see environmental positions be part of the management "rotation," with operational management spending time managing environmental issues and environmental managers finding themselves running a facility.

This convergence tends to demystify environmental issues, taking them out of the special "black box" in which they have resided for decades, and welcomes them to the regular family of important operational issues to be effectively managed in the ordinary course. One should make no more attempt to answer the question "how are we managing our environmental risks today" without an effective system in place than one would answer the question "did we make money this quarter and what is our projection for the next" without numbers produced by an effective and reliable financial and accounting system. In any event, one need not look far afield to find a range of available tools to successfully identify and manage environmental risks and opportunities.

The idea that well-established management tools can be used to successfully manage environmental risks and opportunities does not mean that environmental issues lose their unique character. While environmental risks are not necessarily more complex than other core challenges faced by organizations (and are frequently less complex), they can still have unique characteristics that differentiate them from the latter. The complexity, density, and nature of the legal requirements, the potentially severe consequences of detected noncompliance, the reputational risks, and the degree and extent of stakeholder interest and engagement are all factors that recommend fine-tuning management tools if environmental risks are to be successfully managed. For example, the strict liability/no fault regime of many environmental statutes and the retroactive nature of some of them (e.g., CERCLA), the size of potential penalties for noncompliance even in situations where there may have not been an actual release of hazardous substances into the environment, and the difficult-to-predict media and public reactions to what might objectively be considered low-risk situations create challenges for data-driven management systems. Further, the technical, legal, and social nature of environmental risks may sometimes recommend a more deliberate approach that may run afoul of operational cultures based on speed, agility, or a "bias for action." Simply applying "lean manufacturing," "operational excellence," or "Six Sigma" to environmental risks without tailoring those tools to the unique aspects of environmental risks and opportunities may lead to disappointing surprises and performance.

6.3 Enterprise Risk Management—A Quick Tour

Within the general topic of organizational management, enterprise risk management (ERM) is an umbrella term for a widely used technique to identify, assess, and manage risk on an enterprise-wide basis, typically in the context of one of the overall management or operational systems discussed in the previous section. ERM (and its variants) is an effort to impose discipline on the always-difficult task of identifying and prioritizing risks and establishing effective controls on the risks so identified. Since many organizations use some form of ERM, and environmental issues can and do pose risks, it makes sense for those charged with managing environmental risk to be aware of ERM. It makes equal sense for those managing ERM programs to be aware of the place and relevance of environmental risk. Given that the outputs of ERM processes are frequently evaluated and acted upon at the C-suite and Board levels, it can be an invaluable process from an environmental perspective, while the absence of environmental matters from ERM processes can lead to inadequate communications about and control of environmental risk issues at the highest levels of organizations.

While there are many ERM models, the one that has perhaps received the most attention was created by the Committee of Sponsoring Organizations of the Treadway Commission (COSO) and first published in 1992 as the *Internal Control—Integrated Framework*. Its prominence was firmly established in 2003, when the Securities and Exchange Commission (SEC) decided that the COSO framework could be used as an evaluation framework for management's annual internal control evaluation and disclosure requirements under the Sarbanes-Oxley Act of 2002 (SEC 2003).

The COSO framework was initially aimed at and widely used to manage financial risks and controls. COSO subsequently broadened the scope of its work, developing ERM guidance beginning in 2009. The basic concept of ERM is for organizations to take an enterprise-wide portfolio view of risk, using a disciplined approach to identifying risk at operating units and evaluating and understanding the relationship among the risks. The 2017 COSO ERM Framework has 5 elements (with 20 sub-elements):

- Governance and Culture, which focuses on the organization's risk culture, setting the "tone" for the organization, etc.
- Strategy and Objective Setting, addressing overall strategy, objectives, risk appetites, etc.
- Performance, identifying and evaluating risks and opportunities (both internal and external), and then establishing risk responses in light of the risk portfolio
- Review and Revision, assessing the performance of the entity and the function of ERM, with an eye toward identifying and executing improvements
- Information, Communication, and Reporting, through which relevant risk information flows up, down, across, and through the organization

These components are consistent with the plan-do-check-act framework of most management systems.

The COSO documents have typically focused on financial and business risks, though there was nothing inherent in them that prevented their use for non-financial purposes, such as addressing environmental risks, social issues, etc. However, environmental risks were not always transparently or fully addressed in the ERM systems of organizations. This may have been, at least in larger organizations, because environmental risks might not have easily risen to the level of enterprise-threatening financial materiality necessary to elevate them to a prominent place on the ERM risk grid. This could create asymmetrical situations where, even for companies that might identify environmental protection as a high priority and as central to organizational culture, key environmental issues might nonetheless either be identified as a low priority or be entirely absent from the ERM risk grid shown to C-suite management or the Board.

It is hardly surprising then that the most visible frameworks used to manage environmental risks, such as ISO 14001, were first developed in the 1990s with little reference to COSO and ERM. Since then, not unexpectedly, ISO has entered the ERM space and developed several risk management documents, the most prominent being ISO 31000:2018. At a high level, there are many similarities between the COSO framework and ISO 31000, though COSO still reflects its origins in financial controls, accounting, and audit, while ISO 31000 is shorter and more generic, and perhaps provides more detail on risk management itself and includes the theme of managing and creating value as well as dealing with risk.

It is also not unusual, even in sophisticated companies, to find environmental risk management and compliance programs that have no or little relationship to the companies' ERM programs. This can have unfortunate consequences if ERM is an organization's central tool for communicating and managing enterprise risk at the C-suite and Board levels, yet environmental is on the outside looking in. A major environmental risk might only appear on the radar after the fact, leaving senior management (and frequently the regulators, media, and other stakeholders) to wonder, "What were they thinking?" On the other hand, given the complexity of some organization's ERM systems, managing environmental risks outside ERM may provide for more flexible and nimble (dare we say, "agile") approaches, so long as this does not signal that environmental and sustainability are externalities to the business or reduce their visibility and importance to senior management. This is sometimes done by having environmental risks and opportunities identified, communicated, and managed through a parallel line that still makes its way to the C-suites and the Board.

COSO has worked to explicitly include more non-financial risks into its ERM framework. In October 2018, in conjunction with the World Business Council for Sustainable Development, COSO published *Enterprise Risk Management—Applying Enterprise Risk Management to Environmental, Social and Governance-Related Risks* (COSO ESG Framework). The COSO ESG Framework is an effort to apply their ERM framework to manage environmental, social, and governance risks. The global consumer products company Unilever is used as an example in the COSO ESG Framework, highlighting a broad governance and sustainability strategy that considers a wide range of issues, including agricultural sourcing, nutrition, climate change, deforestation, human rights, fair compensation, data security and privacy, taxes and economic contribution, animal testing/welfare, and packaging/waste, along with ethics, values, and culture. Within this inclusive ESG framework, Unilever manages its environmental issues through an EMS based on ISO 14001. The Unilever example shows that the COSO framework, including the ESG framework, can be used to develop an integrated approach to managing environmental risks and opportunities. One should expect that these risk management tools will continue to grow in sophistication as organizations face the challenges of managing complex risks such as climate change and pandemics.

References

Adkins, L. *Coaching Agile Teams: A Companion for ScrumMasters, Agile Coaches, and Project Managers in Transition*. Boston: Addison-Wesley Professional, 2010.

Business Roundtable. "Statement on the Purpose of a Corporation." August 19, 2019. Accessed November 23, 2019. https://www.opportunity.businessroundtable.org/commitment/.

Chandler, A.D. *The Visible Hand: The Managerial Revolution in American Business.* Cambridge, MA: Harvard University Press, 1977.

Charan, R. and Yang, J. *The Amazon Management System*, Ideapress Publishing, 2019.

Committee of Sponsoring Organizations of the Treadway Commission (COSO). *Enterprise Risk Management—Applying Enterprise Risk Management to Environmental, Social and Governance-Related Risks*, 2018. www.coso.org.

COSO. *Enterprise Risk Management: Integrating Strategy with Performance*, 2017. www .coso.org.

Deming, W.E. *The Essential Deming: Leadership Principles from the Father of Quality*, edited by Joyce Orsini. New York: McGraw-Hill, 2012.

Drucker, P.F. *Management: Tasks, Responsibilities, Practices.* 2nd ed. New York: Harper Business, 1993.

Drucker, P.F. *The Essential Drucker.* New York: Harper Business, 2001.

Elkington, J. *Green Swans, The Coming Boom in Regenerative Capitalism.* Fast Company Press, 2020.

Elkington, J. and J. Zeitz. *The Breakthrough Challenge: 10 Ways to Connect Today's Profits with Tomorrow's Bottom Line.* Jossey-Bass, 2014.

ISO Handbook – The Integrated Use of Management System Standards (ISO, Second Ed., 2018-11).

Kanigel, R. *The One Best Way: Frederick Winslow Taylor and the Enigma of Efficiency.* Cambridge, MA: MIT Press, 2005.

Ohno, T. *Toyota Production System.* Portland, OR: Productivity Press, 1988.

Miller, A. *Redefining Operational Excellence: New Strategies for Maximizing Performance and Profits Across the Organization.* New York: AMACOM, 2014.

Pande, P.S., R.P. Neuman, and R.R. Cavanagh. *The Six Sigma Way: How GE, Motorola, and Other Top Companies are Honing Their Performance.* New York: McGraw-Hill, 2000.

Securities and Exchange Commission (SEC). *Management's Report on Internal Control Over Financial Reporting and Certification of Disclosure in Exchange Act Periodic Reports*, SEC Release No. 34-47986, June 5, 2003.

Webster, K. *The Circular Economy: A Wealth of Flows*, 2nd ed. Cowes, UK: Ellen Macarthur Foundation Publishing, 2017.

7

Compliance and Ethics Programs

Christopher L. Bell

7.1 Introduction

Along with operational/business management systems and enterprise risk management (ERM), compliance and ethics programs play an important role in managing environmental risk. While operations/management systems theory has traditionally had its origins in the business side of the house, and ERM has its origins largely in financial controls, ethics and compliance programs—at least in the U.S.—have their origins in legal compliance. It is not unusual for larger organizations to have all three layers, developed at different times with different "champions" in the organization, which can create challenges for transparent, efficient, and effective risk management.

There are several reasons why a book on managing environmental risk has a chapter on ethics and compliance programs. First, legal compliance is a major component of managing environmental risks. Much of the pressure on organizations to "internalize" environmental issues came from the development and enforcement of comprehensive environmental laws. Second, ethics plays an ever-increasing role in managing environmental risk, reflected in the prominence of sustainability and social responsibility issues, often expressed as environmental and social governance, or ESG. Organizations are coming to realize—in the face of global issues such as climate change, pandemics, access to potable water, the proliferation of discarded micro-plastics, and the ever-growing engagement of the public, shareholders, and other stakeholders—that environmental issues cannot be successfully addressed from purely technical or legal perspectives. Third, governmental authorities, particularly in the U.S., have made effective ethics and compliance programs a central pillar of their evaluation of organizations in the enforcement and regulatory contexts. If something is very important to powerful governmental authorities, *ipso facto* it becomes (or should become) important to organizations operating within their jurisdiction. For all of these reasons, many organizations have developed and implemented general ethics and compliance programs, making it prudent to consider whether and how to integrate or coordinate environmental risk management activities with them.

Ethics and compliance programs are aimed at promoting a culture of ethical conduct, including preventing, detecting, and correcting violations of the law. This framework is well suited to addressing environmental risks, since legal compliance and ethical concepts associated with sustainability are central to successfully managing environmental issues. It is important for those responsible for managing environmental risk to understand compliance and ethics systems because (1) their organization is likely to already have such a system and (2) such systems are of interest to government enforcement authorities and the judiciary, making it prudent to understand what these key stakeholders expect to see. Oddly enough, it is not unusual to see environmental issues managed outside of an organization's formal ethics and compliance program. Indeed, sometimes ethics and compliance programs are very narrowly drawn, managing only the organization's code of conduct and perhaps the anonymous "hotline," leaving most of the substantive compliance work to be done by individual functions in comparative silos. This approach can decrease the effectiveness of the ethics and compliance program and raise questions about the environmental program as well.

The concept of an effective ethics and compliance program is straightforward: does the organization have the leadership, attitude, knowledge, tools, and resources to help its members do the right thing every time? All it takes is one person, one incident, to expose an organization to significant risks and liabilities. If an organization has a good culture and systems in place, bad acts and "bad luck" are less likely to occur. The organization must also have the resources and procedures in place to detect and correct potential problems (and mitigate the consequences) before they get out of control (i.e., no "unfunded mandates").

Since the U.S. Sentencing Commission's Sentencing Guidelines were first published in 1991 and revised several times since then, the public (including shareholders), enforcement authorities, regulators, and the courts have increased their scrutiny and expectations of compliance and ethics systems (U.S. Sentencing Commission, 2018). Organizations characterized by good governance ask:

> Are we meeting all of our major obligations today, and how do we know? Are there some ticking time bombs out there that we won't know about until we read about them in the newspaper? Do we really know what our risks and obligations are (as well as those over the horizon)? Does everyone know what they need to do to make sure we are doing the right thing to stay out of trouble?

These questions could each be expanded to include related social and sustainability issues (or obligations) that look beyond immediate legal requirements or risks. This interrogation may also be understood positively. It need not just have the avoidance of noncompliance or the mitigation of risk as its objective; it can also be aimed at affirmatively meeting challenges and taking advantage of opportunities. For example, in the case of environmental matters, one can use sustainability strategies to develop products and services that are differentiated based on positive environmental outcomes, or actively engage stakeholders rather than be reactive. ESG has become a mainstream concept, used both internally and by external stakeholders to evaluate organizations. For example, the United Nations–supported Principles for Responsible Investment (see www.unpri.org) are used by more than 2,300 signatory institutional investors with over $70 billion under investment to take ESG issues into account when making investment decisions.

These questions (or some form of them) could come from a manager, the Board of Directors, employees, shareholders, investors, customers, the media, governmental authorities, or any number of other stakeholders. An effective ethics and compliance system is a good foundation for producing credible answers to these reasonable questions, just as a good financial and accounting system is necessary to produce consistently accurate and reliable financial performance data. Of course, the function of the system is to not merely answer questions about whether an organization is doing good; it is to enable the organization to consistently produce the conduct and results necessary to produce positive answers (i.e., that it is doing good).

7.2 Why Implement a Compliance and Ethics Program?

7.2.1 Achieve Compliant and Ethical Conduct

The fundamental purpose of an effective ethics and compliance program is to establish and promote a culture and implement the systems that enable and demand that an organization's people do the right thing all the time. To put it negatively, an effective program must prevent, detect, and correct illegal or unethical conduct. An effective ethics and compliance program is an integral component of prudent risk management that decreases the likelihood of conduct that may expose the organization to economic or reputational losses or legal liabilities and promptly corrects and mitigates losses should such misconduct occur. Effective programs should also enhance an organization's capability to be sustainably successful. Promoting and achieving ethical and compliant conduct is part of being a "good corporate citizen" and staying in business.

Elements of effective ethics and compliance programs are increasingly required by law. For example, the U.S. Securities and Exchange Commission has established requirements to implement the Sarbanes-Oxley Act of 2002, such as a code of ethics for senior financial officers, internal accounting controls, and "whistleblower" programs. Indeed, the 2004 revisions to the Sentencing Guidelines, which increased the emphasis on ethics and board responsibilities, were triggered by a provision in Sarbanes-Oxley requiring the Sentencing Commission to review and improve the Guidelines. Further, securities exchanges such as the New York Stock Exchange and NASDAQ require the listed companies to implement Codes of Business Conduct and Ethics.

In addition to these general legal pressures, there are many subject-matter-specific legislative and regulatory requirements (or guidelines) aimed at ethics and compliance programs in areas as varied as pharmaceuticals, health care, import/export controls, bribery and corruption, money laundering, employment law, and government contracting. Implementing effective compliance programs is thus not simply a good idea; it is increasingly the law. In addition, there is a range of nongovernmental guidance documents on this topic, ranging from the work of Committee of Sponsoring Organizations of the Treadway Commission (COSO) on ERM to the environmental management systems and sustainable development documents published by ISO (e.g., ISO 14001:2015 and ISO 26000:2010). ISO has also completed a guidance document on compliance management systems (ISO 19600:2014) and published a standard for anti-corruption management systems (ISO 37001:2016).

The compliance and ethics program envisioned by the Sentencing Commission can include and be integrated with existing controls created pursuant to environmental and sustainability concerns, Sarbanes-Oxley, securities exchange listing mandates, anti-corruption, and other requirements. Indeed, it may be preferable to avoid having multiple, and possibly inconsistent or even conflicting, compliance programs. Further, compliance and ethics programs are often more effective if they are well integrated into everyday business systems and practices rather than being free-standing "black boxes" that can create the unfortunate impression that ethics and compliance are not part of day-to-day conduct.

Those charged with managing environmental risk should be aware that their organization may already have in place several compliance and ethics programs aimed at various substantive topics. These may be operating in silos or integrated into a single umbrella ethics and compliance program. Since they reflect the varied ways in which the organization is already managing risk and compliance issues, it is prudent to understand how and where they work, as well as what is working well and what is not. This will increase the likelihood of success and acceptance of any environmental risk management program.

7.2.2 Satisfy Fiduciary Duties

The importance of compliance programs was highlighted in the seminal *Caremark* case, in which the Delaware Court of Chancery (one of the most influential courts on U.S. corporate law) concluded that individual members of Boards of Directors have a fiduciary duty to ensure that a corporation acts in compliance with applicable laws and has a compliance program to meet this obligation (*In re Caremark Inc. Derivative Litigation*, 698 A.2d 959 (Del. 1996)). In the *Caremark* case, the court expressly discussed the Sentencing Guidelines, noting that

> any rational person attempting in good faith to meet an organizational governance responsibility would be bound to take into account this development [the criteria for compliance programs in the Sentencing Guidelines] and the enhanced penalties and the opportunities for reduced sanctions that it offers.

The Delaware Supreme Court has since confirmed the reasoning in *Caremark*, holding that directors may be liable for failure to exercise appropriate oversight if the directors failed utterly to implement any reporting system or controls or, having implemented such a system, consciously failed to monitor or oversee its operations and thus prevented themselves from being informed of risks or problems requiring their attention. (See, e.g., *Stone v. Ritter*, 911 A.2d 362 (Del. 2006); *In Re Walt Disney Co. Deriv. Litig.*, 906 A.2d 27 (Del. 2006).) These cases have been criticized for not imposing a sufficiently positive duty to engage in more affirmative oversight, and it has been difficult for plaintiffs to successfully plead *Caremark* claims with sufficient particularity to avoid motions to dismiss. (See., e.g., Hill and McDonnell 2013.)

Though recognizing that the burden of pleading a *Caremark* claim was "onerous," the Delaware Supreme Court recently allowed such a claim to go forward, concluding that

> [w]hen a plaintiff can plead an inference that a board has undertaken no efforts to make sure it is informed of a compliance issue intrinsically critical to the company's business operation, then that supports an inference that the board has not made the good faith effort that *Caremark* requires. (*Marchand v. Barnhill*, 212 A.3d 805, 822 (Del. 2019))

Emphasizing that this is a unique board-level responsibility, the *Marchand* court rejected the argument that the requirement for compliance oversight was satisfied by management and operational-level compliance manuals and procedures (*Id*. at 823). A subsequent decision by a Delaware Chancery court allowing a *Caremark* claim to proceed emphasized that a company not only had to have a board-level compliance program but also had to actively implement it (*In Re Clovis Oncology, Inc.*, C.A. No. 2017-0222-JRS (Del. Ch. Oct. 1, 2019)). The *Clovis* court held that, while the company had a compliance program in place, the plaintiff adequately pled that the Board ignored a series of "red flags" regarding the company's compliance with regulatory requirements and internal protocols for the development of drugs. The *Clovis* court also distinguished between the broad discretion and flexibility allowed for board oversight of business risks and the Board's obligation to oversee compliance with "positive law": "As *Marchand* makes clear, when a company operates in an environment where externally imposed regulations govern its 'mission critical' operations, the board's oversight function must be more rigorously exercised." The *Clovis* court went on to quote the Delaware Supreme Court in *Marchand* that this "entails a sensitivity to 'compliance issue[s] intrinsically critical to the company,'" having observed that at certain times the company's Board had appeared to have "hands on their ears to muffle the alarms."

The *Marchand* and *Clovis* decisions clarify that Boards of Directors have a fiduciary duty to (1) ensure that there is an effective compliance program in place and (2) affirmatively engage in the oversight of the company's activities through that program, including paying attention to "mission critical" compliance issues. The Board does not satisfy this fiduciary obligation simply by having a formal program in place but ultimately ignoring the "red flags" of noncompliance waved before it, or by relying solely on management and operational-level programs and systems.

7.2.3 Mitigate the Effect of Noncompliance

Even a good program is unlikely to be 100% effective. Therefore, it is important to be able to detect and fix problems when they are small, and hopefully keep them small. Waiting until a problem is "big enough" to deserve close attention by senior management may produce a regular diet of big problems.

An effective program can also be instrumental in mitigating the enforcement response if wrongdoing comes to the government's attention. The U.S. Department of Justice (DOJ) has developed policies that consider an organization's compliance assurance systems in determining whether and how to respond to noncompliance (see DOJ 2018). DOJ also has subject-matter-specific enforcement policies that delve into the effectiveness of compliance programs, including one on the Foreign Corrupt Practices Act (FCPA) in Section 9-47.120 of the *Justice Manual* .

In 2019 DOJ published more detailed guidance on how it will evaluate corporate compliance systems when making charging decisions (DOJ 2019). DOJ's basic questions are commonsense and straightforward, but it can take a lot of work to be able to credibly answer them:

1. Is the corporation's compliance program well designed?
2. Is the program being applied earnestly and in good faith? In other words, is the program being implemented effectively?
3. Does the corporation's compliance program work in practice?

(DOJ 2019, 1)

DOJ's policies on what constitutes an effective compliance program are consistent with the elements set forth in the Sentencing Guidelines and are directly relevant to environmental risk programs because DOJ plays a significant role in enforcing environmental law and leading the government's legal response to major incidents. The effectiveness of compliance programs is also taken into account by many other Federal agencies in determining whether and what enforcement response to take, including U.S. EPA, the Securities and Exchange Commission (SEC; on financial, FCPA, and "conflict minerals" matters), Department of Commerce (e.g., for import/export programs), Department of Health and Human Services, and the Food and Drug Administration. Many federal agencies (including U.S. EPA) have their own compliance program policies, in addition to DOJ's umbrella policy and the Sentencing Guidelines criteria. Agencies that have no apparent relevance to environmental risk may nonetheless provide useful guidance from a structural or systems perspective. For example, the Department of Treasury included in its *A Framework for OFAC Compliance Commitments* (May 2, 2019) a "top ten" list of root causes of compliance system failures, many of which are easily applicable to environmental systems.

Therefore, DOJ and many other federal (and state) agencies, including environmental regulators, frequently ask for information about an organization's compliance programs when conducting civil and criminal investigations. Demonstrating the effectiveness and credibility of these programs can be an important factor in determining the government's enforcement response. Many organizations invite this inquiry by relying on the "bad apple" defense, in which a company argues that it has good compliance systems and that the alleged violations occurred only because of the bad acts of a few mistaken or rogue individuals. The government will expect a company raising this defense to be able to show that it has an effective program and that the offending individual's conduct was indeed unusual and proscribed.

A corollary to the "bad apple" defense is the "bad luck" gambit. In the environmental context, this frequently arises when there are incidents that occur during natural disasters, such as fires, floods, etc. Such claims are increasingly being met with skepticism, as the public and governmental authorities expect the organizations to have assessed the risk associated with the possibility of natural disasters and to have taken reasonable measures to prevent and mitigate adverse impacts that could arise from such events. At the macro level, there is increasing pressure on organizations to make specific plans for extreme weather events that may be related to climate change. Claiming that a major storm was a "once in a lifetime" event will not be persuasive when it is the third such storm in the past five years, and claims that a facility met a regulatory requirement to plan for 25-year storms might not be accepted by stakeholders in the face of a series of 100-year or 500-year storms. The failure to demonstrate and execute such systemic planning may subject organizations to tort liability, enforcement, and even criminal prosecution for environmental incidents arising out of what, in earlier times, may have been excused as being caused by "acts of God." One can expect similar pressures with respect to organizations ability to prepare for and respond to other public health challenges, such as pandemics.

This is not an abstract concern. The *DOJ 2019 Guidance* emphasizes that the government is not interested in mere "paper programs." The government wants proof that the programs are working (noting that it is primarily interested in the status of compliance programs *before* the alleged offense). For example, if the alleged "bad apple" repeatedly received big bonuses for generating big profits, was regularly promoted, and had excellent performance reviews, investigators will not be inclined to accept the company's position that it consistently enforced the compliance program and may conclude that the company "looked the other way" in the interests of profit and keeping its top producers happy. Similarly, an organization that did not fully evaluate risks associated with extreme weather conditions and did not implement redundant systems to address those risks might not be able to make a credible "bad luck" or "act of God" argument. DOJ addressed the "bad apple" issue directly in Deputy Attorney General Sally Yates' September 9, 2015, memorandum addressing "Individual Accountability for Corporate Wrongdoing."

In matters where the government believes there may be significant civil or criminal misconduct, investigators may review documents and records and interview employees, contractors, suppliers, and vendors, and anyone else they think may have relevant information about the design and implementation of an organization's compliance programs. Simply producing a code of conduct and applicable policies and procedures is typically not enough. DOJ is very clear on this point:

> Even a well-designed compliance program may be unsuccessful in practice if implementation is lax or ineffective. Prosecutors are instructed to probe specifically whether a compliance program is a "paper program" or one "implemented, reviewed, and revised, as appropriate, in an effective manner."

(DOJ 2019, 9)

The goal of these programs, of course, is to prevent and detect misconduct in the first instance and thus avoid serious, including criminal, enforcement consequences. However, organizations that have implemented effective compliance programs may nonetheless find themselves criminally convicted. Convicted organizations may, if they have met the Sentencing Guidelines, benefit from a significant downward adjustment of their criminal penalty. Courts may also require convicted organizations, as a condition of probation, to implement a compliance assurance program (an increasingly common phenomenon).

Depending on the nature of the offense, having an effective compliance program can, under the point system used by the Guidelines, reduce corporate sentences by up to 95%. To receive full "credit," the effective compliance program must be accompanied by self-reporting and full "cooperation" with the government during the investigation (a topic that raises a variety of hotly contested issues, including the extent to which a company must waive any potentially applicable privileges). An organization that has not implemented an effective program will not receive this benefit, and it is not unusual for the court to require the implementation (or enhancement) of a compliance program as a condition of probation. In addition, implementing comprehensive ethics and compliance programs is frequently a condition of so-called "debarment agreements" that allow organizations who would otherwise be barred from contracting with the government due to criminal misconduct to avoid that potentially existential consequence so long as they comply with the requirements of the agreement (see, e.g., EPA, Suspension and Debarment program, www.epa.gov/grants/suspension-and-debarment program). Companies that are either convicted of or plead guilty to violations of environmental laws may find themselves facing mandatory ethics and compliance obligations imposed through comprehensive probation and debarment agreements.

The mitigation benefits of an effective compliance program can also apply to civil as well as criminal situations, where enforcement authorities have significant discretion in the size of the penalties they seek and the burdens of the injunctive relief they would impose. For example, U.S. EPA has "penalty policies" that guide its penalty calculations under most of the major environmental statutes (see www.epa.gov/enforcement/enforcement-policy-gu idance-publications). In addition to considering the gravity of the alleged violation and the extent of actual or potential harm, these policies give U.S. EPA the discretion to give "credit" (i.e., a discount from the maximum penalties allowed by statute) for the organization's good faith efforts to comply or cooperate with the authorities. A credible demonstration that an organization has an effective compliance program can be helpful in obtaining such "credit." In addition, U.S. EPA's penalty policies frequently allow for additional credit for early and voluntary self-disclosure of potential violations. Organizations with effective ethics and compliance programs may be more likely to detect and report issues early enough to take advantage of these opportunities. These policies are sometimes adjusted to address environmental or public health crises such as hurricanes or pandemics (EPA, 2020).

In addition to enforcement policies, U.S. EPA and many states have audit and voluntary disclosure programs that, provided that certain conditions are met (which in the case of some state programs includes prior notification to the authorities), can result in significant penalty reductions or even no penalties at all for violations discovered by audits or compliance programs and promptly disclosed (see, e.g., EPA 2018 and www.epa.gov/compliance /epas-audit-policy). These options are more likely to be available for organizations that have effective programs in place that include good auditing, preventive/corrective action, and internal communications procedures.

However, implementing an effective compliance management system does not guarantee that the government will not press an enforcement case, or that it will immunize a company against criminal liability for the actions of its employees. (See, e.g., *U.S. v. Ionia Mgmt. S.A.*, 555 F.3d 303 (2d Cir. 2009). (In affirming the conviction of a shipping company under the Act to Prevent Pollution From Ships, 33 U.S.C. § 1901, the court noted that compliance programs do not immunize a corporation from liability when employees, acting within the scope of their authority, violate the law, citing *U.S. v. Twentieth Century Fox Film Corp.*, 882 F.2d 656 (2d. Cir. 1989).)

The Sentencing Guidelines' criteria on compliance programs, DOJ's emphasis on compliance programs, and the subject-matter-specific guidance documents issued by many regulatory agencies have had a major influence on prosecutorial, regulatory, and private sector views on compliance and risk management programs and have encouraged their widespread implementation. Accordingly, those charged with managing environmental risk should understand and take into account these governmental criteria, in addition to well-established operational management and ERM frameworks. It would not be prudent to manage environmental risk without incorporating what the government believes are necessary components of an effective program.

7.3 Elements of an Effective Compliance and Ethics Program

7.3.1 Compliance, Ethics, and Culture

The 1991 Sentencing Guidelines originally focused solely on effective programs to prevent and detect criminal violations. The basic goals of an effective compliance program were: (1) reduce the likelihood of illegal conduct occurring in the first place; (2) increase the likelihood of early detection and mitigation should illegal conduct occur; and (3) improve the likelihood of reasonable enforcement response. An effective compliance program may also contribute to other equally important benefits such as improved risk management, morale, productivity, reputation, and shareholder confidence.

The Sentencing Guidelines were significantly revised in 2004, an outgrowth of the spate of financial scandals and the resulting public and regulatory interest in what can be done to increase organizational compliance with the law, particularly at senior management levels (Hill et al., 2013). In addition to preventing and detecting criminal conduct, organizations now had to "otherwise promote an organizational culture that encourages ethical conduct and a commitment to compliance with the law" (U.S. Sentencing Commission, 2018). This change in tone pushed companies to increase the emphasis on culture and ethics in their compliance programs.

This shift reflected an increasing belief that simply writing detailed compliance procedures and demanding that employees follow them (which, ironically, reproduces the "command and control" approach often complained about when embodied in law) was not enough. Rather, an effective compliance program should be built on a culture and ethic of "doing the right thing" as a matter of course in every aspect of personal and organizational conduct. The cliché that "ethics is what people do when they think no one is watching" has a point; no matter how good a compliance program is, it cannot cover everything or be everywhere. So, the growing emphasis on ethics, not just compliance, is to create a general foundation, or culture, of good conduct which, in turn, should also make compliance much more likely. This is particularly applicable to environmental issues, where the focus on the extremely detailed regulatory requirements can sometimes obscure more holistic and effective risk management strategies.

This approach has been described as focusing on the positive "thou shalt," not just the negative "thou shalt not." It is difficult to have a rule for every situation where misconduct could occur. Further, a strictly rule-based program can result in clever individuals attempting to excuse improper conduct by claiming that there were no rules that explicitly prohibited precisely what they did (adopting the theory of "what is not prohibited is permitted"). Sometimes, the more detailed the rules, the more opportunities there are for creative individuals to find opportunities outside the rules. A rules-based approach can result in a "what can (or can't) we do?" perspective rather than a values-based "what *should* we do?" approach. This, in turn, can lead regulators, media, and the public to ask, "what were they thinking?" when the problems begin to pile up. Further, the time-honored public statement that "we fully complied with all applicable legal requirements" is rarely convincing and might be interpreted as conceding that perhaps the "right thing" had not been done.

Commentary to the Guidelines states that management bears a direct responsibility to foster a culture of ethical and compliant conduct. Managers can do this in a variety of ways, including setting a personal example of exemplary conduct, demanding the "high road" from their subordinates, and not accepting "close call" behavior in the interests of short-term financial performance. This requires that ethics and compliance be considered in all of an organization's major systems and programs. For example, it is challenging to expect managers to "do the right thing" at the expense of short-term financial performance if their performance and rewards are measured primarily in terms of that very same short-term performance.

Ethical issues can be raised in everyday business situations, such as asking the tough questions when a subordinate comes in with a "deal too good to be true," or when they suggest that an environmental issue should not be a significant problem as they have an understanding with the regulators or have been told that a requirement will not be enforced. Managers should actively engage, ask the tough questions, and not assume that their subordinates will "report by exception" or conclude that if they do not hear that anything is wrong, then nothing is wrong. There is a big difference between saying the "door is always open if you have any concerns" and actively and regularly raising questions and inquiring about specific compliance issues.

If superiors only grudgingly follow the rules, do not openly and regularly demonstrate ethical conduct personally, and never ask about compliance or ethics, subordinates may conclude that it is not a high priority.

The emphasis on culture has long been accepted as central to an effective occupational health and safety strategy. Though occupational safety regulations are no less comprehensive or detailed than environmental requirements, it is generally accepted that a rules-based approach, by itself, will not achieve superior safety performance. Instead, leading organizations, consistent with (even if typically separately from) the Sentencing Guidelines, recognize that the rules will not be followed unless there is a visible and credible culture, from the top down and bottom up, that embeds safe behavior in everyday conduct. Despite various efforts to "green" organizations and claim environmental protection or sustainability as a core value, environmental still lags behind safety in effectively embedding itself into the culture and everyday conduct of most organizations. Perhaps this is in part because safety is viewed by many as inherently more "personal," even though environmental protection, natural resource conservation, etc., could all be taken just as personally outside the workplace as a matter of nonoccupational safety.

7.3.2 Expanded Scope: Industry Practice and Supply Chain Management

The Guidelines observe that an organization's "failure to incorporate and follow applicable industry practice or the standards called for in any applicable governmental regulation weighs against the finding of an effective compliance and ethics program." This signaled a significant expansion beyond compliance with law to programs that take "industry practice" into account. Companies that have not adopted "best practices" in their industry sector might find themselves having to explain why such practices have not been systematically implemented, even if they are not required by law. Further, those companies that claim to have adopted best practices or voluntary standards may face difficulties if these commitments are not kept. This is a simple but important concept: "Say what you do and do what you say." If the organization makes public commitments that it does not keep, then employees, shareholders, the public, and enforcers may have doubts that the organization is really committed to doing the right thing. This is something that organizations making broad public claims regarding sustainability, or generating well-produced "sustainability" reports, should keep in mind: does an organization's "ground game" match its public claims?

The Commission also believes that the scope of larger organizations' ethics and compliance programs should extend beyond organizational boundaries. The Commission, in commentary, suggests that larger organizations, where appropriate, should encourage smaller organizations, particularly those that seek a business relationship with the larger organization, to implement effective programs. This is consistent with the developing interest in "social responsibility" and "sustainable development," pursuant to which organizations are taking steps to address a wide range of issues (e.g., working conditions, wages, bribery and corruption, and environmental protection) in their relationships with suppliers and contractors.

There is an increasing expectation among stakeholders such as public interest groups, shareholders, investors, and customers that organizations must be able to credibly demonstrate that their entire business footprint, throughout the value chain (i.e., from raw materials and suppliers through distribution and end-of-life of products), is being operated in a sustainable and responsible manner. This is not solely an issue of "doing good"; an organization may suffer significant economic and reputational harm and possibly incur legal liability for the misdeeds of agents, distributors, suppliers, or contractors thousands of miles away. The massive increase in transnational FCPA enforcement, Federal Acquisition Regulations (FAR) requiring that federal contractors and subcontractors implement comprehensive ethics and compliance programs, requirements for traceability and reporting associated with "conflict minerals," the supply chain controls necessary to comply with chemical regulatory requirements such as the Toxic Substances Control Act (TSCA) in the U.S. or the EU Regulation on the Registration, Evaluation, Authorization and Restriction of Chemicals (REACH) and California's "Transparency in Supply Chains Act of 2012" all reflect the expansion of this trend into enforceable legal requirements.

The *DOJ 2019 Guidance* also emphasizes supply chain issues, devoting an entire section to "third party management." DOJ expects compliance programs to include third parties and vendors to be considered when assessing the risk of noncompliance, developing operational controls, engaging in mergers and acquisitions, and auditing and conducting root cause analyses of incidents (DOJ 2019, 6–8). This renewed emphasis on ethics and compliance programs aimed at third parties has a definite place in environmental risk management systems.

7.3.3 Elements of an Effective Program

7.3.3.1 Leadership and Oversight

The Guidelines establish three senior levels of responsibility for compliance programs: (1) "governing authority"; (2) "high-level personnel"; and (3) "substantial authority personnel."

A. The Board

The "governing authority" is the Board of Directors (or its equivalent). The Guidelines establish the following key requirements for the Board:

- The Board "shall be knowledgeable about the content and operation of the program to prevent and detect violations of the law."

- The Board "shall exercise reasonable oversight with respect to the implementation and effectiveness of the program."
- The Board (or a Board committee) is expected to periodically receive information on the implementation and effectiveness of the program.
- The Board must receive training regarding the operation of the program and the Board's responsibilities in it.

Therefore, the Guidelines describe an explicit responsibility on the Board of Directors (or other governing authority, if there is no Board) to know how the compliance assurance system works and to oversee its implementation. While the Guidelines do not specify how the Board should meet these obligations, the oversight function is often assigned to the Auditing Committee. Some organizations, seeking to increase the focus on ethics and compliance and recognizing the already-increased burdens on Auditing Committees due to Sarbanes-Oxley, have established new Ethics or Compliance and Ethics Committees. Committees have also been established to focus on specific programs, such as Board-level Environmental, Health and Safety Committees.

However Board oversight is organized, the Board should review and approve the key elements of the ethics and compliance program and exercise regular and visible oversight. The Board can do this in at least two ways. First, the Board should integrate ethics and compliance issues in the exercise of its regular duties. For example, when business strategies are being reviewed or senior management compensation approved, the Board should be asking the "is this the right thing to do" questions. Ethics and compliance should not be reserved solely for formal meetings arranged for that purpose. In addition, the Board (or some part of it) should regularly review the ethical and compliance performance of the company, including the operation of the program. The Board should also establish procedures for prompt communication about serious ethical or compliance issues.

Separate from the Guidelines, the *Caremark* line of cases emanating from the Delaware courts are a reminder of the common-law fiduciary duty of Boards to actively monitor and oversee the activities of the organization through a defined compliance system.

B. Senior and Middle Management

The Guidelines also establish responsibilities for senior management. All management, not just those formally charged with compliance, must be knowledgeable about the operation of the ethics and compliance program, perform their duties in a manner designed to prevent and detect violations of the law, and promote an organizational culture that is committed to compliance and ethical conduct. This has sometimes been described as the "tone at the top," though it is far more than "tone" and indicates the day-to-day actions of leaders. DOJ has also emphasized this point:

> The company's top leaders—the board of directors and executives—set the tone for the rest of the company. Prosecutors should examine the extent to which senior management have clearly articulated the company's ethical standards, conveyed and disseminated them in clear and unambiguous terms, and demonstrated rigorous adherence by example. Prosecutors should also examine how middle management, in turn, have reinforced those standards and encouraged employees to abide by them. See U.S.S.G. § 8B2.1(b)(2)(A)-(C).

(DOJ 2019, 9)

This is one of the most important elements of an effective ethics and compliance program and any environmental risk management program that is part of it. It is common sense that what the leadership and management of an organization regularly express is important to them and is what will dictate the path of the organization. If leadership's daily oversight focuses on production rates and margins, that is what the rest of the team will focus on as well. Management that does not include ethical, compliant, responsible and sustainable conduct as part of the expectations they convey in their everyday activities sends a direct message to the organization as to the relative importance of those issues. If the only person talking about ethics and compliance is the compliance or ethics officer, that may send the message that compliance is a marginal staff priority, not a core operational issue. For example, inviting the ethics officer to come to talk to a department about compliance is not as effective as having senior operational management being directly and regularly engaged with ethics and compliance issues.

This does not mean that ethics and compliance should displace other core issues such as production, quality, innovation, etc. Rather, it is to suggest that when organizations are engaged in the daily effort to be successful, part of that daily effort should be to achieve success in an ethical and compliant manner. The best-designed and sophisticated ethics and compliance programs, including environmental risk management programs, that do not include the constant engagement of management at all levels will always face very difficult implementation headwinds. On the other hand, even modest systems that are driven by an engaged management are more likely to be effective, sustainable, and capable of continual improvement. It is usually relatively easy to correct or improve a defective procedure or element of a system; however, it is extremely challenging, if not impossible, to overcome a management team that is not engaged with ethics and compliance issues.

This concept is embedded in several of the more popular business management models (discussed in Chapter 6). For example, "operational excellence" emphasizes a culture of management leadership and direct and continuous engagement. Integrating ethics and compliance issues, including environmental risk, into such systems can be effective. There are organizations that have integrated environmental risk management and compliance into their daily "OpEx" conference calls involving senior and operational management, in which any environmental incidents that occurred in the past 24 hours are discussed, evaluated, and next steps, if any, agreed upon. Leaders and management at all levels can make ethics and compliance part of the organization's everyday conversation, not an externality that is grudgingly tolerated. Leadership on ethics and compliance is also demonstrated by having a positive day-to-day attitude and emphasis. Leaders who regularly complain about legal requirements or treat legal requirements or internal ethics policies as unfortunate barriers to business planning and success are sending a message that ethical and compliant conduct is not highly valued. Superior occupational health and safety performance is also often achieved in organizations with an embedded "safety culture" where safety professionals play a supporting, not dominant, role.

The importance of resources available to support ethics and compliance cannot be overestimated. These resources are typically under the control of operational and financial management, not the individuals in charge of administering the ethics and compliance program. If people do not have time to do their jobs, and performance goals continue to climb, the temptation to cut corners on "soft" issues such as compliance and ethics grows. This observation does not apply just to the compliance function itself, but to the entire organization. A common discovery during internal investigations is that "we did not have time to worry about [whatever the compliance issue was] because we were too short on staff after the cutbacks," or "management attention to compliance and ethics really dropped off after the big reorganization and the drop in stock price, after which 100% of our effort had to be focused on increasing profit and market share" (or dealing with the debt on which the deal was based). Another issue frequently discovered during post-incident root cause analyses is that a long series of apparently modest cuts/reductions, whether in personnel, budget, maintenance, etc., any of which by themselves did not raise a risk flag, taken together were a significant contribution to a failure (CSB, 2007). If compliance and ethics are treated as something people must pay attention to in their spare time, you will get a "spare time" program.

The Guidelines also require that the organization

> use reasonable efforts not to include within the substantial authority personnel of the organization any individual whom the organization knew, or should have known through the exercise of due diligence, has engaged in illegal activities or other conduct inconsistent with an effective compliance and ethics program.

"Substantial authority personnel" include senior managers, as well as individuals who exercise substantial supervisory authority (e.g., a plant manager or a sales manager). There is not an absolute bar on hiring individuals with a history of misconduct into positions of responsibility. The organization should consider how the individual's record of misconduct relates to the specific responsibilities the individual will be assigned to, how recent the misconduct was, and the nature of other misconduct that might not be directly related to the individual's anticipated duties. But this is a factor that will be considered if an organization raises the "bad apple" defense.

C. Compliance Officer(s) and Personnel

"High-level personnel" (i.e., senior management) must ensure that the organization has an effective compliance and ethics program. "Specific individual(s)" within senior management must be assigned overall responsibility for the program. While the Guidelines do not use the term "compliance officer" or "ethics officer," the person who is assigned this responsibility often carries such a title. Ethics or compliance officers can report directly to the CEO or the Board (or both), and are sometimes folded into the legal function and report to the general counsel. This is what DOJ has to say on the matter:

> Effective implementation also requires those charged with a compliance program's day-to-day oversight to act with adequate authority and stature. As a threshold matter, prosecutors should evaluate how the compliance program is structured. Additionally, prosecutors should address the sufficiency of the personnel and resources within the compliance function, in particular, whether those responsible for compliance have: (1) sufficient seniority within the organization; (2) sufficient resources, namely, staff to effectively undertake the requisite auditing, documentation, and analysis; and (3) sufficient autonomy from management, such as direct access to the board of directors or the board's audit committee. The sufficiency of each factor, however, will depend on the size, structure, and risk profile of the particular company.

> **(DOJ 2019, 10)**

There are many approaches to organizing the compliance function, with no single best answer or practice. The path ultimately chosen will depend on a variety of factors, including organizational culture, structure, and lines of communication. One approach is to appoint a senior operating manager as compliance and ethics officer, sending the clear message that ethics and compliance are core business values. There may be a concern that this puts the fox in charge of the chickens, but if that is indeed the case, the organization has significant problems that might not be solved by devolving the job to "independent" staff.

Another approach is to put an "ethics professional" in that role, potentially emphasizing "independence" and "ombudsman" themes, though this can unintentionally create the impression that ethics is a professional staff responsibility divorced from normal business operations or compliance. Some organizations make the general counsel the compliance officer, though concerns have been raised that this can impose an overly legalistic tone to the program and might also discourage internal communication given the reluctance of some to talk to lawyers. Another option is to have a "compliance committee" composed of executive management that oversees the operation and performance of the compliance system (which typically includes the compliance officer). Of course, it is important to avoid the error of implying that the staff compliance function is directly responsible for compliance at the operational level; that responsibility belongs to the managers and employees in the operations.

Day-to-day responsibilities for the compliance program may be delegated to levels below the senior manager who is the compliance officer. For example, drafting procedures, conducting training and audits, operating the ethics hotline, or collecting and summarizing performance information may be done by staff reporting to the compliance officer. The senior compliance function may be supported by a cross-functional compliance staff team (e.g., human resources, legal, finance, audit, environmental) that assists in the day-to-day operation of the compliance program. Some of the work might also be delegated outside the compliance function. For example, some companies delegate the investigation of noncritical hotline calls to their human resources department. Companies may also retain third-party contractors to manage their 24-hour hotlines. Again, it is important to not confuse responsibilities for the administration of the ethics and compliance program with the responsibility for ethical and compliant conduct.

Persons to whom such day-to-day compliance responsibilities have been delegated must be provided adequate resources and authority to do their jobs, and must also report periodically to senior management on the effectiveness of the program. The Commission emphasized that while various duties may be delegated, the ultimate responsibility for the effectiveness of the compliance program resides with senior management.

The Commission commented that if the compliance officer does not have day-to-day responsibilities for the program, then the person(s) with the day-to-day responsibilities should report directly to the Board, or a Board committee, at least annually on the effectiveness of the program. The concept here is that senior management and the Board (or a committee thereof) should periodically hear from people "in the trenches" about how the compliance program is really working. This is necessary for the Board to meet its oversight responsibilities.

The Commission's 2010 amendments encourage companies to allow individuals with operational responsibilities for the compliance program to be able to report directly to the Board by offering penalty reductions if certain conditions are met (including that the violation was discovered by the compliance and ethics program before it would have been detected outside the organization, compliance and ethics personnel had nothing to do with the violation (including not being "willfully ignorant"), and the violation was reported to appropriate government authorities). The direct reporting to the Board contemplated by the Commission includes giving these compliance personnel express authority to communicate directly with the Board on any matter involving actual or potential criminal conduct, and no less than annually on the implementation and effectiveness of the program.

This is an example of distinguishing between responsibility and accountability for the ethical and compliant conduct itself and an independent line of communication for verifying and communicating about such conduct. Giving the compliance officer an independent and direct line of communication to the Board on ethics and compliance issues contributes significantly to the effectiveness of the system and the Board's ability to meet its oversight obligations, but does not (nor should not) transfer the responsibility or accountability for ethical and compliant conduct to the compliance officer.

7.3.3.2 Risk Assessment

Based on the principle of "you can only manage what you know," the Guidelines provide that organizations "shall periodically assess the risk of criminal conduct and shall take appropriate steps to design, implement or modify each requirement … [in the balance of the Guidelines] … to reduce the risk of criminal conduct identified through this process." In the commentary, the Commission states that organizations should assess the nature and seriousness of potential violations of the law, the likelihood that certain violations of the law may occur because of the nature of the organization's business, and the prior history of the organization. Assessments of legal risk are not a one-time event; they must be conducted on an ongoing or periodic basis so that the assessment is kept current and the compliance system tailored to address the current risks. This is a point of emphasis for DOJ:

> The starting point for a prosecutor's evaluation of whether a company has a well-designed compliance program is to understand the company's business from a commercial perspective, how the company has identified, assessed, and defined its risk profile, and the degree to which the program devotes appropriate scrutiny and resources to the spectrum of risks.
>
> **(DOJ 2019, 2)**

The design and implementation of the ethics and compliance program must consider the risks identified in the assessment. The commentary to the Guidelines provides examples of how this is expected to work. For example, organizations that employ sales personnel who have flexibility to set prices "shall establish compliance standards and procedures designed to prevent and detect price fixing," while organizations whose sales employees have the flexibility to "represent the material characteristics of products shall establish compliance standards and procedures to prevent fraud." This general principle has obvious application on the environmental front: well-run organizations should (must?) know and manage their environmental risk footprint.

Once the risks are identified, organizations must develop and implement programs that effectively address those risks (including practical procedures that direct employees, in language that they can understand and that is relevant to their responsibilities, on what they must do to comply). Compliance programs that generate detailed memoranda formally describing generic legal requirements without reference to the company's risks and activities are not likely to be effective or credible. How the organization tailors its attention and resources based on the degree and nature of risk is also relevant. For example, devoting significant resources to managing what the organization has identified as low-level risks may be questioned (DOJ 2019, 3).

There is no consensus on what constitutes the "best practice" for conducting risk assessments, and perhaps this is as it should be. At one end of the scale are extremely detailed and quasi-quantitative assessments, while at the other end some companies identify their primary legal challenges in relatively broad strokes. Another issue is the extent to which any risk assessment of ethics and compliance issues should be part of or integrated with an organization's ERM program (assuming it has one). The benefits of having "one-stop shopping" for risk identification and prioritization are obvious, including built-in visibility at the Board and C-suite levels and identifying, evaluating, and managing ethics and compliance (including environmental risks) at the same table as the rest of the organization's core issues. Care must be taken to ensure that the ERM system is properly designed to identify and evaluate risks that might not be easily quantifiable or may be of high consequence but very low probability. Such risks (e.g., an environmental risk with relatively low direct financial consequences but immense reputational consequences) might not rise to visible levels in ERM systems that are largely financially driven. Sometimes organizations modify their ERM systems to address environmental and sustainability risks in a supplement or even have a completely different process for evaluating them. The last approach can have the disadvantage of not taking full advantage of the resources and experience of the ERM system and requires care to be sure that the outputs get the same high-level attention.

There is no requirement in the Guidelines for complex quantitative assessments or ranking systems; indeed, in some situations such approaches may be unnecessarily expensive and cumbersome and not provide significant marginal value. For example, a company that does business outside the U.S. should be able to determine reasonably quickly, if it does not already know, that compliance with chemical or product stewardship requirements in target countries is a high-priority issue (though identifying the precise contours of that exposure can be complex). The potential applicability of non-U.S. legal requirements must also be considered (e.g., sellers of chemicals and consumer products must address a rapidly proliferating and global collection of product registration, content, labeling, and performance requirements).

There can be a tendency for risk assessments to focus largely on the risks of legal noncompliance and possibly overlook other issues that may yet not have a specific legal consequence. More forward-looking companies expand their risk assessments beyond purely immediately applicable legal requirements and consider risks associated with potential future developments, reputational risks, issues of concern to stakeholders important to the success of the organization, etc. These organizations may include ethical and reputational risks in their assessments, even if there might not be any specific associated compliance risk. One example is evaluating the potential impacts and opportunities associated with climate change by organizations whose own emissions of greenhouse gas might not be directly regulated or that might face significant climate-related risks even if their own operations might not generate significant greenhouse gas emissions. Incorporating ESG into risk evaluation may increase the focus on value chain issues such as labor and public health conditions. The increased emphasis on ethics has also generated an interest in evaluating organizational culture, seeking information on issues such as the "tone" set by management, employees' views on the relationship between ethical and financial issues, and the ease with which legal or ethical concerns can be raised. This is consistent with the Guidelines; if senior management is responsible for building an organizational culture that supports compliant and ethical conduct, it needs to know what that culture is.

7.3.3.3 Standards and Procedures

The organization must have standards and procedures reasonably capable of reducing the prospect of misconduct, addressing the results of the risk assessment. This is a commonsense requirement; if the Board and senior management expect people to do the right thing, it is important to communicate, through standards and procedures, what and how to do the right thing. It is also difficult to fairly enforce the program if employees have not been given information on what they are expected to do. Most organizations include procedures aimed at reducing the likelihood of noncompliance and conduct that infringes on organizational rules, and at preventing and mitigating identified risk.

At the highest level, direction is typically provided through a simple one-page policy expressing the organization's commitment to ethical and compliant conduct. Some organizations will layer additional one-page policies behind the overarching policy, addressing specific topics such as environmental, anti-corruption, etc. However, developing a multiplicity of top management policies over time can create complexities and confusion, as it is easy for the "messaging" in each policy to be slightly different. The volume of messages is sometimes increased with "mission" and "values" statements. The policy is just a beginning point that by itself does not comprise a system and will not accomplish much.

The policy is normally backed up with a code of conduct that summarizes the key requirements and expectations of the organization. The code of conduct typically summarizes the organization's key obligations and risks and, in a few short paragraphs, the related expectations. For example, a code of conduct may have sections on antitrust, corruption, environmental, health and safety, political contributions, diversity, and harassment, etc. Information on communicating questions or concerns to management (including anonymously), as well as discipline for violating the code, are in almost every code. Codes may have electronic links to more detailed or specific procedures.

At the next level down, procedures should provide practical direction aimed at making sure that people know what to do on an everyday basis. Procedures are intended to go beyond high-level summaries contained in codes of conduct or short statements in policy or mission statements, which rarely provide the specific direction or detail required for the task. Procedures should be evaluated from the perspective of the ultimate user, not the "experts" who write procedures. Detailed procedures that may be legally correct are often not effective because either the procedures do not fit the practical day-to-day context of employees or employees simply cannot understand them (or they are just too long). For example, long, dense memos written by lawyers on Clean Air Act compliance are not likely to be read by busy power plant operators.

One approach is to build compliance points into regular business practices, such as triggering questions about environmental permitting on construction or planning documents, or questions about the regulatory status and risk profile of raw material inputs during the design process. These procedures may extend beyond the normal internal operations of the organization. For example, in the case of product stewardship, procedures should include the appropriate management of environmental risk issues throughout the value chain (e.g., supplier selection, natural resource uses, and chemical content of supplied materials). Where business processes are computerized, mandatory electronic compliance "checkoffs" can be added. This approach can reinforce compliance in an everyday manner, rather than relying solely on annual training that someone is then expected to remember for the rest of the year or periodic auditing. Building compliance into normal business documents also creates records of compliance that can be audited.

Procedures supplement, but do not replace, the foundation of having a culture of ethical conduct. One cannot have a rule covering every conceivable situation, and the absence of a specific rule should not be viewed as permission to act. Ethics has been loosely described as what you do when no one is watching. That is why the Guidelines address ethical, not just compliant, conduct. When in doubt, one would want members to be asking themselves not just what is compliant conduct, but also what is the right thing to do. Could you explain your conduct to your family? A reporter? How would this look on the front page of the newspaper? The legally defensible action may not be the correct action. Asking these questions in advance can avoid the post-crisis question that is universally asked after questionable behavior has been disclosed or exposed: "What were they thinking?"

In addition, the Board and senior management play a critical role in establishing the overall tone of the company. A management ethic that always rewards financial performance without asking how that performance was achieved is at risk. It is rarely convincing, after the fact, to attempt to evade organizational responsibility for the actions of employees by claiming that they did not follow company procedures if the company never tried to find out if anyone was paying attention to the procedures.

7.3.3.4 Communication and Training

The Board, senior management, and indeed all management have an obligation under the Guidelines to promote an organizational culture that encourages ethical and compliant conduct. Regularly communicating about the importance of good conduct is one of the most important ways, in addition to their actions, that the Board and managers can build an ethical culture. At its most effective, this communication takes place as an integrated element of regular business interactions, rather than solely as special "ethics" or "compliance" events (though the latter can play a role in an internal communications program).

Training is also a critical element of internal communications. The Guideline's training requirements explicitly cover the Board and all management, including senior management. This commonsense provision may cause significant changes in many organizations' training programs. Training is often thought of as something that production workers or contractors must go through, not managers (particularly senior managers). Even in companies that have extensive new-employee training and qualification procedures, it is common to find relatively little training at most levels of management, particularly on compliance-related issues.

The objective of effective training programs is straightforward: members need to understand the organization's expectations regarding ethical and compliant conduct, their roles and responsibilities regarding such conduct, and when and where to get help. It is particularly important that the training be relevant to the participants' work life; generic training is rarely effective. This can be an issue with computer-based training (CBT), which sometimes is provided by third-party vendors and may not be tailored to the specific circumstances of the organization or audience. CBT no doubt is a cost-effective method of disseminating a large volume of information to many people, but the value of interactive face-to-face sessions should not be ignored.

Further, training should not be viewed as a replacement for effective leadership in which management demonstrates on a daily basis the importance of ethics and compliance and ethical conduct is embedded in organizational culture. One sometimes runs into situations where senior management is puzzled and disappointed that misconduct occurred, saying, "We give them ethics training every year, what else are we supposed to do?" Similarly, an annual "certification" that members have read the code of conduct only goes so far. Training is only one part of an effective compliance program, supplementing the heart of an effective compliance and ethics program, which is a culture that visibly and tangibly builds good conduct into everyday business life. If the only time employees hear about compliance and ethics is during annual training sessions sponsored by the compliance office, it is possible, perhaps even likely, that this topic will be viewed as being of marginal importance.

Though the Guidelines do not mandate external communication about these issues, leading companies are increasing their levels of communication about ethics and compliance, with a trend among major multinationals to publicize regular reports on social responsibility or sustainable development (frequently guided by criteria such as those developed by the Global Reporting Initiative (GRI) or the Sustainable Accounting Standards Board (SABS) and encouraged by organizations such as the World Council on Sustainable Development (www.wbcsd.org), CERES (www.ceres.org) or documents such as ISO 26000). Many companies are posting their code of conduct on their external websites. Of course, it is important that the public communications are backed up with effective systems and good performance. An annual sustainable development report should be the outcome of a sustainable development strategy (i.e., a report is neither a strategy nor a system).

7.3.3.5 Monitoring and Auditing

Organizations must take reasonable steps to ensure that the compliance and ethics program is being implemented and followed. In addition to the daily engagement of management and members in the implementation of the system, the building blocks to accomplish this are monitoring and auditing to detect illegal conduct, and periodic evaluations for the effectiveness of the compliance and ethics program.

A strong auditing program is crucial to demonstrating to regulators and enforcers (or other stakeholders) that the organization has a credible ethics and compliance program. If an organization is not looking hard at its own conduct, the government will assume that the organization does not take the program seriously and is not interested in detecting problems. It is almost always better for the organization to find a problem before anyone else does, or before a whistleblower or ex-employee broadcasts it. Auditing also demonstrates to employees and others subject to the program that the organization is indeed taking the commitment to compliant and ethical conduct seriously. If you are not looking closely, why should they pay close attention?

The compliance auditing program should, at a minimum, cover the legal issues identified in the risk assessment. In a comprehensive program, the risks may include issues not directly related to specific compliance requirements and should also be within the scope of the audit. In this context, organizations should consider the failure to comply with the organization's requirements and procedures to be "findings" that must be corrected. The credibility of programs and procedures is significantly diminished if the failure to conform is not a "real finding" unless it can be tied to a specific regulatory requirement.

The compliance and ethics program itself must also be audited. This typically involves (1) verifying that the elements of the program are in place and (2) obtaining objective evidence through interviews, document reviews, and observations that the program is being implemented. For example, program audits would look beyond whether noncompliance with environmental requirements has occurred and assess the existence and effectiveness of environmental procedures, management leadership on such issues, etc. These audits can often catch potential issues before they turn into real problems because they focus on the extent to which members understand what they are supposed to do and whether they are doing it, and not on the narrower issue of whether anyone has yet to step over the line. Sometimes the most important findings in an audit are systemic issues that have not yet ripened into noncompliance or incidents. In the event questionable activities have occurred, the review would include systems issues such as determining why the potential misconduct occurred and why it was not detected and reported.

Program reviews increasingly include assessments of "corporate culture." These reviews explore what managers, employees, contractors, and others think about the organization, what motivates them, what are the "unwritten rules" on how "things really work," what they believe is acceptable and unacceptable behavior, whether they feel comfortable raising concerns or questions, whether they feel pressures to get the job done and cut corners, etc. Such reviews can also help understand what is being heard by employees, which might be very different from the message that management thinks it is sending. Again, this information can be very useful in identifying and correcting situations where unethical or illegal conduct might be likely and prevent it before it occurs.

"Corporate" audit programs can frequently be a resource for environmental risk systems. Corporate audit typically employs full-time professional auditors who are regularly out in the field, putting them in a position to include some environmental issues in their scope. While the corporate audit team might not always have legal or technical environmental expertise, they can be trained to keep an eye out for specific issues and are typically well placed to evaluate the effectiveness of systems and procedures.

The auditing procedures must be accompanied by corrective and preventive action procedures aimed at determining the cause of any detected shortcomings, fixing the problems, mitigating the consequences, and preventing their recurrence (including changes to the ethics and compliance program). This is the "fix what you find" part of the process.

The corrective and preventive action procedures should be triggered by detections of noncompliant or unethical conduct from any source, not just auditing. Government investigators pay close attention to the corrective action process. It is very difficult to credibly explain detected but uncorrected problems. Indeed, knowingly allowing misconduct to continue can be the launching point for a serious criminal investigation.

It is also important to have defined processes for internal and external reporting. The Board (or the Auditing Committee) and senior management should receive regular reports on audit results and the status of corrective action. As has been well publicized in many corporate scandal cases, the government and public are not receptive to the "I did not know" defense by senior managers. The Board and senior management need to know how the ethics and compliance program is performing, including information such as the results of audits, the status of corrective actions, summaries of hotline reports, etc. This information should be presented in a meaningful way so that the key points and trends are identified and appropriate actions can be taken. For example, simply electronically copying senior management on detailed audit reports is usually not effective. The Board (or Auditing Committee) has a particularly important oversight role where there is a possibility that senior management might be implicated (including separately retaining counsel to conduct independent investigations if necessary). Internal reporting can also be integrated into regular business communications, recognizing that there are trade-offs between immediate and effective communications and the potential loss of whatever privileges and confidentiality protections that may attach to audit information, depending on how they are designed.

Practices regarding required or prudential reporting to the authorities should also be developed. Failure to comply with mandatory reporting requirements can have very serious consequences. Even where there are no mandatory reporting requirements, regulators, enforcers, and courts may take a company's voluntary disclosures and cooperation into account in determining what, if any, enforcement response is appropriate. As noted at the beginning of this chapter, one mitigation value of an effective compliance program to avoid indictment or a severe sentence is also tied to voluntary reporting and full cooperation with the government (with the caveat regarding the extent to which the government can demand the disclosure of privileged information as a condition of cooperation, a complex topic on its own). Organizations should also consider U.S. EPA's auditing and voluntary disclosure programs and the state equivalents. There are a variety of criteria that must be considered to benefit from these enforcement discretion programs, including a relatively short amount of time to submit voluntary disclosures, time frames to complete corrective action, and, in some states, having to give the regulators prior notice that the audit is being conducted. (EPA 2001S.)

7.3.3.6 *Internal Reporting and Investigation*

Organizations must have internal reporting and communications procedures that allow employees and agents to report potential or actual misconduct anonymously or confidentially and be protected against retaliation for such reports. These procedures should also be available to those seeking guidance about ethical or compliance issues, not just those who have complaints. This element in the Guidelines is essentially a broader version of the whistleblower provisions of Sarbanes-Oxley that are aimed at internal reporting of allegations regarding financial and accounting matters. Confidential/anonymous internal reporting is meant to supplement, not replace, the normal lines of communication in a company.

An effective internal reporting structure should include investigation, tracking, follow-up, and corrective, mitigation, and preventive procedures. The 2010 amendments to the Sentencing Guidelines included new application notes regarding corrective and preventive action. The factors that the Commission states organizations should consider include voluntary reporting, restitution to victims, and a review and possible changes to the program to prevent future occurrences (including the possibility of bringing in a third party to review the adequacy of any modifications and their implementation). Organizations must consider how to structure internal investigations, including when and whether to bring in third parties, to conduct effective and perhaps equally importantly, credible, investigations in the event of reported alleged misconduct.

Failure to quickly investigate and resolve concerns can decrease morale and the credibility of the compliance program and allow small problems to quickly escalate into legal, economic, and reputational crises. It is particularly important to have trained internal investigators to respond to complaints or concerns, combined with a mechanism to alert counsel when it is necessary to conduct a privileged investigation or, in the case of serious matters, engage outside counsel to conduct investigations.

The investigation and tracking procedures should apply to all complaints/ concerns that are raised, whether they are made directly to supervisors or through the "hotline." Companies that formally track only "hotline" calls may not realize that they are leaving out considerable volumes of data because more complaints were made to supervisors than through the "hotline."

An effective and credible internal reporting mechanism must be well publicized and easy to use. Otherwise, it will create the impression that the company does not really want to receive complaints or questions. Simply putting information about the program on the company intranet is usually not sufficient. Posters, training, and other means of regularly reminding employees and agents about how to raise questions or concerns are recommended. Even the program's name should be carefully thought through: "hotline," "helpline," "whistleblower line," and "ethics line" all convey slightly different messages.

Whistleblower requirements (including protection for whistleblowers) continue to proliferate, with the SEC's 2011 rules implementing Section 922 of the Dodd-Frank Act as an example. Multinationals face challenges implementing uniform internal reporting programs that meet U.S. requirements (e.g., Sarbanes-Oxley Act, Dodd-Frank Act, and the Sentencing Guidelines) while not running afoul of non-U.S. requirements (e.g., EU privacy and labor law requirements, along with the EU's comprehensive General Data Protection Regulation). Non-U.S. privacy and employment laws can also create challenges in conducting U.S.-style investigations involving interviews of employees and the review of records.

7.3.3.7 Discipline and Incentives

The Guidelines direct that the compliance and ethics program be consistently enforced, and that appropriate disciplinary measures be taken when someone engages in illegal conduct or does not follow the requirements of the program. This is very important to the credibility of the program. If individuals who violate the law or internal policies/procedures are not disciplined, that sends the message that management does not really take compliance and ethics seriously. This can become particularly problematic when a stellar performer from an immediate financial perspective is perceived by his or her colleagues as "getting away with" violations of company procedures. The failure to impose discipline is also an important signal to regulators and enforcers about management's commitment to compliance. If the "bad apple" is getting bonuses or just getting away with it, it makes it more difficult to argue that the individual was really acting outside of company policy.

The Guidelines expand the scope of discipline beyond those who engage in the actual misconduct to those who fail to take reasonable steps to prevent or detect misconduct. In other words, a manager who "looks the other way" or who fails to implement the compliance program's measures to detect misconduct should be subject to discipline. This provision reflects the skepticism that greets the "I did not know" defense.

The 2004 revisions to the Guidelines added a new component: the program should include incentives to perform in accordance with the company's compliance and ethics program. This is a somewhat controversial provision, since some commentators believe that it sends the wrong message to reward individuals for good conduct (implying that good conduct is above and beyond what is expected), while others believe that this is an important positive message that should be used to counterbalance purely "negative incentives" (i.e., punishment) and to demonstrate that good conduct, not simply financial performance, will be rewarded. Care must also be taken to avoid creating incentives to not report problems (e.g., where reporting misconduct in one's department might create a "negative hit" on one's bonus).

It is important that both discipline and incentives be consistently applied. Failure to do so can have several negative impacts, ranging from decreased morale to allegations of discriminatory practices. Again, implementing this element of a program on an international level must consider potential legal limitations on the ability to impose discipline on employees in the countries in which the organization operates.

7.4 Conclusion

Effective ethics and compliance systems are a well-established fixture in the business landscape and can be part of the foundation for effective systems to manage environmental risk. The Sentencing Commission's Guidelines establish an overall compliance system "umbrella" under which the various other programs applicable to an organization, ranging from Sarbanes-Oxley to money laundering to environmental, may be implemented. Compliance systems can also be integrated with broader management systems aimed at quality, efficiency, enterprise risk management, sustainable development, or social responsibility.

Regulators, enforcers, courts, and other stakeholders expect effective ethics and compliance systems, and prudent companies implement them, not merely to satisfy some formal provision of the Sentencing Guidelines, but also to improve performance and prevent and detect misconduct. The most important measure of an effective compliance system is whether it works, not whether it meets the Guidelines "on paper." In other words, will an organization's system increase the likelihood that its members will consistently do the right thing? This question is at the heart of an effective system to manage the environmental risks that are of significant public and governmental concern.

References

Department of Justice (DOJ). "Sections 9–28.000—Principles of Federal Prosecution of Business Organizations." *Justice Manual* [formerly, *U.S. Attorney's Manual*], 2018. Accessed November 23, 2019. https://www.justice.gov/jm/jm-9-28000-principles-federal-prosecution-business-organizations.

DOJ. *Evaluation of Corporate Compliance Programs—Guidance Document, Updated April 2019.* 2019. https://www.justice.gov/criminal-fraud/page/file/937501/download.

Environmental Protection Agency (EPA). EPA's Audit Policy . Accessed May 10, 2020. https://www.epa.gov/compliance/epas-audit-policy.

EPA. "EPA Announces Renewed Emphasis on Self-Disclosed Violation Policies." May 15, 2018. www.epa.gov/compliance/epas-audit-policy.

EPA. "Enforcement Policy, Guidance & Publications." Accessed November 23, 2019. https://www.epa.gov/enforcement/enforcement-policy-guidance-publications.

EPA. "Suspension and Debarment Program." Accessed November 23, 2019. www.epa.gov/grants/suspension-and-debarment-program.

Hill, C., and B. McDonnell. "Reconsidering Board Oversight After the Financial Crisis." *University of Illinois Law Review* 2013, no. 3 (2013): 859–80.

PRI Association. "Principles for Responsible Investment." Accessed November 23, 2019. www.unpri.org.

U.S. Chemical Safety and Hazard Investigation Board (CSB). *Investigation Report, Refinery Explosion and Fire, BP Texas City*, March 20, 2007. Accessed January 6, 2020. https://www.csb.gov/bp-america-refinery-explosion/

U.S. EPA, COVID-19 Implications for EPA's Enforcement and Compliance Assurance Program, March 26, 2020.

U.S. EPA, State Audit Privilege and Immunity Laws & Self-Disclosure Laws and Policies (updated 5/2015), Accessed May 10, 2020, https://www.epa.gov/compl iance/state-audit-privilege-and-immunity-laws-self-disclosure-laws-and-policies

US Sentencing Commission. "Chapter 8—Sentencing of Organizations." In *Guidelines Manual 2018*, 2018. Accessed November 23, 2019. https://www.ussc.gov/sites/default/files/pdf/guidelines-manual/2018/CHAPTER_8.pdf.

Yates, S.Q. "Individual Accountability for Corporate Wrongdoing." *DOJ Memorandum*, September 9, 2015. https://www.justice.gov/archives/dag/file/769036/download.

8

Environmental Risk Management Systems

Christopher L. Bell

8.1 Introduction

Environmental management systems (EMS) are closely related to, and can be integrated with, the business, risk, and ethics/compliance systems discussed in the previous chapters. (For the purposes of this chapter, EMS will be used synonymously with environmental risk management system.) Further, most of the elements of those frameworks, such as management leadership and review, risk assessment, procedures, measuring performance, and auditing and corrective action, apply with equal force to EMS. All the elements of business or operations management systems, enterprise risk management and effective ethics and compliance programs discussed in the Chapters 6 and 7 should be considered in developing and implementing an EMS. This chapter focuses on environmental issues and does not retrace the steps of the previous chapters.

Some of the elements of business or ethics/compliance programs might appear in slightly different forms in the typical EMS, though organizations have had success sharing elements across different roles. For example, it is not unusual for organizations to have common systems for managing documents and records, and the corporate auditing function may play a role in environmental auditing. On a more substantive level, computerized work and process management tools can be used for financial, production, maintenance, quality, and environmental purposes. Even at senior management levels, the intensive management engagement and review elements of "operational excellence" can easily incorporate an environmental component.

The governmental and stakeholder pressures and expectations regarding more general systems apply with greater force to managing environmental risks. In addition to the general principles established by the Sentencing Commission and DOJ, U.S. EPA has also held forth on systems generally and EMS specifically. EPA issued a position statement encouraging EMS on May 15, 2002 (updated on December 13, 2005), and announced *EPA's Strategy for Determining the Role of Environmental Management Systems in Regulatory Programs* on April 12, 2004. EPA has also "walked the talk," having implemented formal EMS at dozens of its own facilities (frequently based on ISO 14001). (See, generally, https://www.epa.gov/ems.) The governmental uptake of EMS extended well beyond EPA; for example, the Departments of Energy and Defense have each made significant efforts, typically modeled on ISO 14001, to implement EMS. (See, e.g., DOD 2016.) The governmental emphasis on the systemic management of environmental issues has also found its way into the influential force of government procurement, with the inclusion of EMS elements and "sustainable acquisition" in the Federal Acquisition Regulations (FAR). (See, e.g., 76 Fed. Reg. 31395 (May 31, 2011).)

The increased complexity of environmental regulations has implicitly driven organizations to develop sophisticated procedures, if not fully developed systems, to achieve and maintain compliance. For example, the Clean Air Act's Title V permitting requirements and the risk management program requirements drive systematic approaches to compliance. Similarly, comprehensive chemical regulatory frameworks such as REACH in the EU or the Toxic Substances Control Act (TSCA) in the U.S. compel disciplined attention to product development and value chain management.

Public and stakeholder pressure, including from consumers, public interest groups, and litigants, has also increased the challenges of identifying and managing environmental risks. With an increased interest in ESG issues such as climate change, renewable resources, public health (e.g., pandemics), responsible sourcing, socioeconomic conditions of the supply chain, etc., effective environmental management is no longer limited to dealing with regulations and governmental authorities. These ESG issues often go well beyond the law and seek organizational responses on issues that are not regulated at all. These challenges can be magnified by the velocity of social media.

These themes are driving organizations, private and public, to better and more credibly manage an ever-increasing array of complex environmental risks. This is evident in the broad uptake of EMS. For example, as of 2018 approximately 307,000 ISO 14001 certificates were extant globally, covering over 1,180,000 sites, with approximately 12,000 registered sites in North America (ISO, 2018). By the end of 2018, over 14,000 sites were already certified to the ISO 45001 occupational safety management standard that was published in 2018. This is only the tip of the iceberg, since most organizations that implement an EMS do not seek third-party certification, including those that use ISO 14001 as a model. Suffice it to say that it has become the norm for organizations of any significant size or complexity to consciously manage their environmental risks through some form of an EMS.

8.2 Scope and Design

One starting point for evaluating an existing EMS or developing a new or enhanced one is to consider the context within which the organization operates and the potential scope of the system.

Understanding the context of the organization is a commonsense concept that was added to the 2015 edition of ISO 14001. Before getting into the weeds of specific environmental issues, it is useful to identify and understand what the organization does and its place in the world—not just the environmental world, but also the business, economic, and social world. This should also include understanding the various stakeholders with an interest in the organization's activities, who can range from employees and shareholders to neighbors, customers, and the public at large (and, of course, the suite of relevant regulators). The evaluation should include the products and services it delivers, the areas where it operates and the markets where it sells, its broader "value chain" (e.g., raw materials and energy sourcing, suppliers and other vendors, distribution systems, downstream customers, etc.), the business sector(s) within which it operates, etc. A thorough understanding of the organization's overall context makes it more likely that the key environmental risks and opportunities will be identified and less likely (though not a guarantee) that the organization will be blindsided by a major unanticipated issue. The public health crisis associated with the COVID-19 pandemic, along with the crushing economic consequences, will no doubt heighten the scrutiny of global value chains.

This analysis should be able to draw from existing business systems, since many organizations have "mission statements" and business plans, and enterprise risk management-type programs may have already outlined some of the organization's context. In any event, it is prudent to coordinate (or, at a minimum, be aware of) these activities, since it may be counterproductive if the context developed from the environmental perspective is not recognizable to the business. However, the "business context" alone may not be comprehensive. For example, an understandable focus on customers, shareholders, and employees may result in other key contextual points getting less attention, such as upstream supply chain issues or a broader view of influential stakeholders. On the other hand, the organization may have already committed to broad values and goals, such as service to the community or sustainability, that can and should inform the direction of managing environmental risks. For example, an organization that has adopted the Business Roundtable's 2019 "Statement on the Purpose of a Corporation" should take those commitments into account when developing and implementing their EMS.

Understanding the organization's contextual footprint will (or at least should) illuminate the scope of an effective environmental risk management system. For example, the scope of a multinational organization's system will likely differ from that of a single-location entity that has a relatively compact and local value chain (though with the growth of internet-based commerce, even small businesses can have a global footprint). Likewise, organizations in the commercial/industrial space may operate in a context that is quite different than that of public-facing organizations. Having a good grasp of the big picture should enhance a robust and comprehensive evaluation of the organization's risks. This process may also initiate an iterative relationship between managing environmental risk and operational management generally, since it may bring to the foreground issues not previously given much priority.

Whatever context within which an organization operates, many practical decisions must (or should) be made about the overall design of the EMS. Will it be a centralized one-size-fits-all system, will a thousand flowers be allowed to bloom, or something in between (e.g., a consistent overall framework with centralized auditing and oversight, but local flexibility)? For multinationals, will each country or region find its own course? Will the system take a broad approach to risk, or will it focus primarily on compliance with legal requirements? Will it address the full range of the organization's activities, services, and products, or will it retain a traditional emphasis on facilities? How, and to what extent, will the system include the value chain, both upstream and downstream? Will it encompass broad issues such as sustainable development and social responsibility?

Another issue is whether and the extent to which environmental management will be integrated with the organization's existing operating systems. For example, if the organization has built its commercial success around "lean manufacturing," "operational excellence," or "agility," how or to what extent will or should the EMS use or be integrated with those processes? Will environmental risk evaluation be integrated into an existing ERM program? Or will environmental risk be managed largely in its own structure, separate from line management? How these types of questions are answered will make a significant difference to the ultimate design and execution of the system. For example, "operational excellence" already incorporates a vigorous management engagement component into which the typical EMS "management review" element could be integrated. Another common integration issue is the relationship between managing environmental risks and occupational health and safety. The pendulum on this last point swings regularly, as organizations appear to regularly integrate and disintegrate environmental, health and safety (EH&S) management. ISO 14001 and ISO 45001 can be implemented together to create an integrated EH&S management system.

There are no "right" answers to these questions or a single "best in class" approach. As in the case with claims about the newest management fad, assertions that the "best" approach to manage environmental risk has been discovered should be viewed with great caution. Much depends on the organization's culture and how it operates and manages its business. For example, an organization with a highly decentralized culture may not welcome a centralized EMS. On the other hand, a flexible and qualitatively oriented EMS may fall flat in an organization with a centralized approach characterized by an engineer-driven quantitative culture. One can do worse than starting by understanding how an organization is already successfully managing core elements of the business and evaluate whether and how that approach can work for the EMS. However, operational principles might not always be perfectly suited for managing all environmental risks. For example, there may be times when you might hit the pause button on the "bias for action" and make sure you are doing the right thing before proceeding. Further, there are many options on how various components of systems can be mixed and matched. For example, as noted in Chapter 6, Unilever is using the COSO ESG Framework for managing its overall sustainability strategy while using an ISO 14001–based EMS to manage specific environmental issues.

These scoping considerations can raise legal issues. Organizations with subsidiaries or that participate in joint ventures or partnerships should balance the risk of taking the desired direction with an understanding that one might step over a line that would allow "piercing the corporate veil" or create similar liability exposure to the parent organization. Similar liability issues come up in the context of addressing environmental issues in the value chain (e.g., vendors and contractors).

The consequences of the path(s) chosen should be kept in mind. For example, a decision to limit an EMS to legal compliance risks does not mean that risks not directly related to specific compliance requirements go away. Integrating environmental risk into normal business and risk management systems does not mean that expected performance can happen automatically: it will take considerable effort to ensure that the necessary knowledge, procedures, and resources are put into play to address complex environmental issues. Organizations should also look to internal consistency. For example, if the organization makes broad policy announcements regarding sustainability and publishes an annual sustainability report, signs up to an industry code (e.g., the ACC's Responsible Care®), or adopts the Business Roundtable's Statement on the Purpose of a Corporation, it will raise an eyebrow if the EMS does not expressly address those same issues.

How these structural issues are addressed will help inform what guidance or models an organization may want to look to in developing and executing its system. Within the generally accepted principles of the "plan-do-check-act" model, there is much to choose from, including the very popular ISO 14001, the guidance provided by the Sentencing Commission and DOJ, the vast business management literature, and the ever-growing literature on how businesses can manage a range of sustainable development and social responsibility issues (e.g., ISO 26001). There is no "right" answer, so long as the result produces a credible and effective program that produces results. If an organization desires the recognition or discipline of third-party certification (or there are market or legal demands to obtain the certification), then ISO 14001 may be the likely choice (though ISO 14001 can be and is widely used for guidance without seeking certification). However, it cannot be overemphasized that third-party certification to ISO 14001 is not necessary to the design and implementation of an effective EMS, and ISO has published extensive guidance on the flexible implementation of ISO 14001 that is not tied to certification (ISO 14004: 2016 and ISO 14005: 2019).

8.3 Leadership, Roles, and Responsibilities

An effective and engaged leadership is central to any management system, playing a central role in almost every framework from the "operational excellence" model, to the Sentencing Guidelines and DOJ policy, to ISO 14001. There was a time when this was phrased in terms of management "support," sometimes accompanied by "my door is always open." Some decades ago, this might have involved occasional pronouncements from senior management on the importance of environmental issues and proclamations that "we have a great environmental team that has our full support." With the advent of ISO 14001 and its "management review" provisions, this was often translated as formal semi-annual or annual review sessions where the environmental manager would make a presentation on the preceding period's environmental performance and the plans going forward.

Leading management engagement practices go well beyond such passive approaches. For example, under the "operational excellence" model, management at all levels, including senior management, are directly engaged on a daily and weekly basis in the organization's key operational issues, with critical review and continual improvement being an essential element of business. Where environmental is folded into an operational excellence approach, that can lead to direct and close to daily involvement of management in managing environmental risks. Some organizations include environmental, health, and safety issues in their daily morning "OpEx" call, as well as in the regular management of operations. In this context, managers are no longer passive spectators who "support" upon request an EMS run by environmental professionals; rather, they seize the reins of responsibility and action themselves.

Active governance of environmental risk extends to the board level or, as the Sentencing Guidelines put it to encompass organizations that may not have a board, the "governing authority." As reflected in the Sentencing Guidelines and the *Caremark* line of cases on fiduciary duty, effective governance must include the active and systematic oversight by the Board, not just senior management. Strengthening Board and senior executive oversight has been identified as an important step in preventing major environmental and safety incidents (CSB, 2020). Therefore, a well-designed EMS should include provisions for Board monitoring and oversight of environmental risk and compliance management and extend to ESG issues as well (Ramani, 2019).

ISO 14001 is somewhat ambiguous on the degree of board involvement, since it establishes roles for "top management," defined as the "person or group of people who directs and controls an organization at the highest level" (ISO 14001-2015, Clause 3.1.5). This could be interpreted as including the Board or governing authority of the organization, but could be viewed as encompassing only the C-suite, since in many circles "top management" does not include the supervising board. On the other hand, this ambiguity, or flexibility, is consistent with the ability to implement ISO 14001 at the facility or business unit level, even if there is not an organization-wide board-supervised EMS. For example, ISO 14001 was designed to allow it to be used by a single facility, in which case "top management" could be the plant manager.

A corollary to management engagement is management responsibility and accountability for the environmental issues and risks that fall within the ambit of their authority. While environmental professionals (technical and legal) might provide staff assistance and expertise, the ultimate responsibility for managing the risks and related compliance should lie with those responsible for the activities that create the issues in the first instance. Management engagement and participation directing the issues that they own has replaced the more passive concept of "management support" for a staff-driven EMS.

Senior management provides the framework for their engagement and the EMS through the development, approval, and execution of a top-level policy that includes the organization's key commitments on environmental issues. At a minimum, such policies include commitments to comply with the law and the organization's requirements, and typically also include statements related to broader issues such as sustainability and the prevention of pollution (the latter, along with continual improvement, being elements of a policy conforming to ISO 14001). This can be a free-standing policy, or the environmental/sustainability points can be integrated into a broader organizational policy statement.

One challenge in creating such policies is reaching a balance between uplifting and aspirational statements while being realistic about what the organization can achieve. An organization can rapidly lose credibility, both internally and externally, if it makes promises it cannot keep. As is commonly said in the ISO 14001 context, one should say what one does and do what one says. One quick test is to reverse an organization's policy and transform it into a series of assessment questions to see just how realistic the policy is. It is a red flag if the immediate response is "well, we didn't really mean that we were going to fully do XX." Organizations should "walk the talk," not just "talk the talk" (or worse yet, only "talk the walk").

In addition to the vital leadership/engagement roles of management, the important roles and responsibilities in the EMS should be defined and well communicated. Many of these might not attach to environmental professionals. For example, maintenance may have many important responsibilities to maintain pollution control equipment, procurement may review and enforce the qualifications of contractors or vendors, and product designers may be the gatekeepers on environmental aspects of product design. The important point here is that roles and responsibilities in an effective EMS should not be limited to the responsibilities of environmental professionals.

Unfortunately, it is not unusual to find that an organization's "environmental manual" is simply a description of what the environmental professionals do, which is only a small subset of the activities and responsibilities through which an organization successfully identifies and manages environmental risk. Indeed, the environmental responsibilities that are probably most important to delineate are not those of the environmental professionals, who are typically relatively small in number and probably have a pretty good idea as to what they are supposed to do anyway. This is yet another reason why it can be helpful to engage cross-functional teams to develop an EMS, since left to their own devices, environmental professionals may not have the knowledge or authority to populate a comprehensive responsibility and accountability matrix for EMS purposes (sometimes expressed through a "RACI" chart, summarizing who is responsible, accountable, consulted, and informed).

The structure of the EMS itself should also be well defined. There are a variety of approaches, ranging from having a senior vice president of EH&S who has direct access to the Board overseeing the system, to having mid-level environmental professionals in charge and reporting to other functions, such as the general counsel, compliance officer, or human resources (sometimes in the case of occupational health and safety), or putting a senior operational manager in charge. For an EMS to be consistent with the Sentencing Guidelines and DOJ policy, those in charge of the EMS must be able to trace a direct and credible reporting line to senior management and the Board. It is generally accepted that the better practice is for the top defined environmental position to be as senior as possible. This enhances the likelihood that environmental will be at the senior management table and signals that environmental is on a par with other core functions of the organization (e.g., finance). It may also be difficult to explain to employees, governmental authorities, the public, and other stakeholders that environment and sustainability are very important to the organization if the visible positions associated with these issues occupy lower portions of the organizational chart. On the other hand, the positioning of the senior environmental professionals should not be confused with where the responsibility for environmental performance and compliance lies.

One EMS design issue in constant flux is the position and role of environmental professionals. Many organizations rotate between having environmental (or sometimes EH&S) in a centralized corporate headquarters function to devolving environmental professionals out into operations, leaving a much smaller presence at the corporate level. Sometimes these rotations of style can occur so quickly that one transition may not even be fully completed before the course is reversed. Various forces appear to be at play in this process.

One is a widespread concern that environmental must be "independent" of operations for fear that if left under the control of operations, environmental professionals will be pressured to ignore noncompliance and possibly even facilitate it. This perspective suggests that operations are fundamentally adverse to environmental protection and compliance, and that environmental professionals effectively serve as internal environmental police. A strong and centralized independent environmental group can be a very powerful tool to achieve acceptable environmental performance, prevent adverse events, fight for an adequate budget, etc. This strategy is often adopted in the aftermath of significant environmental incidents or enforcement cases, where there is significant pressure to improve environmental performance quickly and a traditional "command and control" structure is familiar to regulatory and enforcement authorities.

This strategy does have limitations. First, it implies that significant responsibility for compliance rests with the environmental professionals rather than operations. Second, it can create an adverse relationship between operations and the environmental professionals and decrease the extent to which environmental expertise is called upon (i.e., the EHS professional as the "police"). Third, it can place most of the environmental expertise in the control and disposition of a central corporate function, making it less flexible and responsive to local operational needs. Also, from an overall management efficiency perspective, large centralized corporate environmental groups can be an easy target during economic downturns or corporate reorganizations.

Embedding environmental professionals in the operations that create and manage environmental risks and compliance has its attractions. This gets them closer to operations so that they better understand what is going on, from environmental, technical, and business perspectives. This allows them to more quickly provide advice and assistance that is better tailored to the specific facts and the needs of the business. It also attaches the cost to operations so that they can better internalize and understand the consequences of their activities, rather than relying on corporate largess. It also provides an opportunity to have a leaner corporate staff, which is frequently popular with senior management and shareholders. Of course, this approach has its limitations as well, the most significant being a concern that this puts the fox in charge of the chickens, since this puts operations in control of managing environmental risk and compliance, which may not be their highest priority. In addition, a fully decentralized approach may result in enterprise-wide environmental risks not being effectively identified or managed, and it can weaken oversight and assistance roles such as identifying and keeping up with legal requirements, training, and auditing.

There are things one can do to navigate between these two poles. One is for the Board and senior management to clearly and explicitly attach responsibility and accountability for environmental compliance and risk to operational management and clarify that environmental professionals have a staff role. This approach, embodied in direct and daily management engagement and participation, as well as direct, immediate, and visible consequences for inadequate environmental performance (just as would be the case for inadequate financial performance), may decrease the likelihood that environmental professionals reporting through the operational chain of command will be pressured to look the other way.

To the extent there are legitimate concerns that the operational side of the house is adverse to environmental protection, that organization may have more serious structural problems than will be solved by adjusting the organization chart of the environmental group. Organizations can also develop hybrid systems where environmental professionals out in the field might have solid line reporting responsibility to operations and dotted line reporting to corporate environmental (or the other way around). This provides field environmental professionals with a "life line," and senior corporate environmental professionals with visibility into how operational environmental issues are being managed.

A corporate environmental presence can be maintained to focus on enterprise-wide issues. For example, corporate group responsibilities can include developing and administering the enterprise-wide EMS, identifying and evaluating risks (particularly those that might not be visible at a specific operational level), collecting and evaluating performance data, and providing support on tracking and advising on developments in legal and other requirements and related advocacy. Lastly, without conceding that there is an adverse relationship between operations and environmental, maintaining the environmental audit/verification process outside of operations is also a good practice (though operations should have their own checking and verification processes and not depend solely on a periodic corporate audit that might occur only once a year or even less frequently).

8.4 Assessing Risk

At the heart of any environmental risk management system or EMS, regardless of the model chosen, is accurately and comprehensively evaluating the organization's environmental risks (or, using the language of ISO 14001, "environmental aspects"). The progression is straightforward: once an organization has identified its key environmental issues, the balance of the EMS is designed around effectively managing them in light of its policy commitments, legal obligations, etc.

Organizations should have a defined process for assessing environmental risks. This will impose discipline on the process, assist the users of the outputs, understand where they came from, and promote consistency of analysis and outputs over time. Having a defined procedure also helps the organization explain what it did in the event that a third party may inquire. For example, DOJ may ask very specific questions about an organization's risk management assessment process:

> Risk Management Process—What methodology has the company used to identify, analyze, and address the particular risks it faces? What

information or metrics has the company collected and used to help detect the type of misconduct in question? How have the information or metrics informed the company's compliance program?

(DOJ 2019, 3)

Organizations that seek ISO 14001 registration must also demonstrate the adequacy and effectiveness of their procedure to identify significant environmental aspects.

In this context, the term "risk assessment" is used in a relatively informal "civilian" manner, and it is not intended to suggest that organizations must conduct an assessment that would meet quantitative scientific risk assessment standards. Many, probably most, environmental risks being evaluated at the organizational level are not easily susceptible to a scientific quantitative risk assessment, nor is such an assessment typically necessary for management purposes. This process is not the equivalent of conducting a site-specific public health risk assessment at a contaminated site. Evaluating the nature, extent, likelihood of occurrence, and the consequences (i.e., gravity) of occurrence of various environmental scenarios typically involves a series of judgment calls for which relevant and reliable data may be limited. For ease of decision making and communication, it may be possible to assign numbers to those judgments and then slice and dice those numbers. However, an environmental risk matrix populated with very precise looking numbers should not be mistaken for a science-driven analysis.

While translating professional judgments about risk into numeric values is popular and can be useful, there is limited value in getting into extended discussions as to the precise numeric values to be assigned. Organizations can waste a lot of time debating whether the likelihood that an event will occur should be scored a "3" or "4," or whether the consequence is a "6" or an "8." It is important that the consumers and users of such quantitative matrixes are aware of the central underlying subjective assumptions and judgments. As one commentator has aptly observed, mathematical models and algorithms are "opinions embedded in mathematics" (O'Neil, 2016). This suggests that a certain skepticism be applied when faced with sharply drawn quantitative evaluations of risk or other quantitative metrics. Sometimes the participants already have a good idea as to what the big risks are and sometimes go back and adjust the scores when an issue of significant intuitive concern scores "low." However, one of the benefits of a disciplined process is that it serves as a check on long-held intuitions or oral traditions and to enhance the likelihood that risks show up on the radar screen that had not received full consideration or analysis. It may also enhance consistency over time.

If an organization already has a risk assessment system in place, such as ERM, the relationship between that system and evaluating environmental risk should be explored. For example, if ERM is viewed as the central process whereby key enterprise risks are communicated to the Board and C-suite, what would it mean to evaluate environmental risks outside of that process? On the other hand, folding environmental into ERM may result in significant environmental issues disappearing into the background in the context of other business risks, raising questions as to whether the ERM methodology is well suited to assess environmental risk. As noted in Chapter 6, COSO has been revising its ERM model to better take ESG issues into account.

Determining the scope of the environmental risk assessment process is a crucial step and is tied to the scope of the overall system, which has already been discussed at length earlier in this chapter. If the scope of the EMS is narrowly drawn, the scope of the risk assessment will likely follow suit. Traditionally, assessments of environmental risk have focused on legal compliance risks associated with facility operations and the risks associated with cleaning up contaminated sites (whether legacy or future), often coinciding with the relatively narrow scope of the typical environmental group and EMS. This approach is not fit for purpose in a more dynamic and broader ESG context of environmental risk management, sustainability, and social responsibility. For example, issues such as climate change or the environmental footprint of an organization's supply chain might not appear in risk assessments tied to facility-specific legal requirements. Further, risks such as product design and related end-of-life issues, natural resource use, human rights, working conditions, public health, mergers, acquisitions and divestitures, or environmental issues associated with the acquisition and use of raw materials and energy can fall through the cracks of a facility-oriented risk assessment.

Apart from direct compliance obligations, the economic and reputational consequences of environmental risks gone wrong continue to increase. These consequences are not limited to headline events such as the Deepwater Horizon incident in the Gulf of Mexico. Significant environmental issues can appear on organizations' doorsteps in unexpected ways from unexpected directions. For example, while the scientific community had been raising concerns about plastics in the environment for quite some time, the issue burst into the public's imagination in 2018 with videos of turtles impaled with plastic straws, expanding into public, and in many cases regulatory, calls for the elimination or reduction of plastics in packaging and other uses. Public-facing companies then faced stakeholder pressure, or the opportunity, to revise their sales and packaging practices to effectively address (or even get ahead of) this sudden attention. Similarly, the COVID-19 crisis has demonstrated the importance of agile and resilient risk management.

Organizations are also recognizing that issues that have long been the foundations of sustainability discussions, such as climate change or access to useful or potable water and other public health issues, can pose direct risks or offer significant opportunities. This can result in environmental risks moving up the priority list, sometimes for sectors that did not traditionally view themselves as having significant environmental footprints. For example, forward-looking real estate development firms are putting issues such as climate resiliency (e.g., rising sea levels and extreme weather events), energy efficiency, distributed energy generation, and renewably sourced materials on their priority list. Organizations in the food production sector are finding that community pressures regarding heavy water use can create existential pressures on their business models apart from any specific regulatory requirements (e.g., the concept of a "social license"). The public health and economic consequences of the COVID-19 pandemic have raised a bewildering array of risk management issues. Thus the evaluation of risks can be accompanied by an assessment not only of how to mitigate those risks but also identifying opportunities arising from those risks, such as transportation solutions with smaller greenhouse gas footprints or consumer products manufactured and designed in line with "circular economy" concepts that consider the full life cycle of the service system.

The risk assessment should also consider the full range of the organization's activities, services, and products. This has traditionally started with the environmental aspects of the organization's facilities. Including products broadens the process to consider everything from the materials used in the manufacture of products to the environmental performance during the use of the product to the product's end-of-life. Evaluating services can bring into play environmental issues such as the greenhouse gas footprint of a purely web-based marketplace that, while it may not have any facilities or trucks of its own, relies on a global transport infrastructure based largely on fossil fuels. Outsourcing has also become a part of the risk evaluation question: shareholders, investors, customers, regulators, and other stakeholders are increasingly asking if (or demanding that) products come from sustainable sources or socially responsible supply chains. Moving an environmentally sensitive operation out of one's own facilities and shifting it up or down the value chain does not eliminate it from the organization's risk footprint (and significant concerns are now being raised about global value chains and public health in light of the COVID-19 pandemic).

The organization's activities should be broadly evaluated to include risks associated with mergers, acquisitions, divestitures, joint ventures, asset dispositions, and retirements, etc. While it has become commonplace for acquiring companies to do extensive due diligence to avoid or mitigate the risks associated with legacy contamination (e.g., "all appropriate inquiries" and Phase 1 environmental assessments associated with the bona fide/innocent purchaser defense provisions of Superfund), the level of due diligence on a target's broader environmental risks is sometimes less thorough (e.g., evaluating a target company's product or chemical portfolio to project potential future regulatory or market barriers and liability risks). Divestitures and asset retirement are also worthy of attention as risk points, given that risks at operations slated for sale or shutdown may fall off the radar screen. Significant environmental liabilities have been generated by incidents at such facilities.

Another element of environmental risk that an effective system should consider is third-party litigation. Environmental incidents, or even just the fear of environmental exposures, can result in litigation brought by neighbors or customers. Traditional common-law concepts such as nuisance are also being creatively applied to issues such as climate change or cleaning up rivers and harbors that have been contaminated by the actions of countless entities and individuals over a period of centuries. These risks are not necessarily directly tied to what one might call scientifically assessed risk. For example, chemicals that have been found to pose little if any significant public health risk by any number of peer-reviewed studies or governmental agencies might nonetheless be subject to aggressive litigation where the outcome is based largely on the effectiveness of advocacy and the decisions of non-scientists (i.e., the jury). Securities litigation has also become an avenue for environmental litigation, with shareholder derivative claims for share losses following major environmental incidents like clockwork. Multinational companies can also face litigation in their "home countries" arising from the environmental consequences of their far-flung operations anywhere in the world. For example, in *Vedanta Resources PLC v. Lungowe* [2019] UKSC 20, the UK Supreme Court allowed environmental statutory and tort claims to be brought in UK court against the UK parent company associated with the copper mining operations of a subsidiary in Zambia, with one of the factors taken into account by the court being the parent company's public statements regarding group-wide sustainability standards and environmental controls.

The economic and reputational impacts of third-party litigation can be significant, even existential. Nevertheless, though litigation exposure based on traditional legal principles has been around for generations and were even the initial legal response to environmental risks, litigation risks are sometimes not taken fully into account in risk assessments. It is unclear why this is the case; perhaps it is because these risks somehow get lost between basic compliance risks and the new attention to sustainability issues, and that litigation is frequently viewed in reactive terms and the province of litigators.

Thus, an effective EMS will have procedures in place to comprehensively and regularly assess its environmental risks. This assessment should consider the full range of risks, from big-picture issues such as climate change and water and raw materials acquisition to product design, performance, and end-of-life to value/supply chain management to facility-level issues. This can be done in a free-standing process or be integrated with ERM or similar frameworks, so long as the right people are at the table to do the job and the resulting information gets to where it needs to go, from the Board and C-suite to the individuals whose daily actions directly affect the successful management of those risks. Having the right people at the table can be a challenge, since many of an organization's environmental risks might not be in the ambit of environmental professionals. For this reason, assessing environmental risk is best viewed as a multi-functional team process that is not limited or perhaps even led by the environmental function. Some organizations have designed environmental risk assessment processes that are driven primarily by members who are directly involved with the activities, products, or services at issue, with environmental professionals bringing their expertise to the table in a staff role. This approach can also facilitate operational "ownership" of and familiarity with their environmental risks, decreasing the likelihood that the outcome of the assessment will be received as a foreign document imposed from the outside by the environmental group.

The environmental risk assessment process normally involves a prioritization process so that resources will be allocated based on some reasonable relationship to the risk. As noted above, this is an inherently imperfect process based on collective judgment and cannot be reduced to a purely objective process ... no matter how much it is dressed up to look like one. There is nothing magical about the building blocks of any risk prioritization process: evaluating the nature, likelihood, and consequences of the issue at question.

The degree of regulation is commonly considered, with the type of enforcement response and possible penalty often part of risk matrix calculations (e.g., issues where noncompliance might result in an administrative notice of violation or have only a very low penalty exposure might have a lower rank). Environmental exposure criteria are also frequently used, with small releases in buildings or soil that are not reportable to the authorities on one end of the scale, and larger spills—ones that are reportable or that cause public or employee exposures—populating the other end of the scale. Some organizations will include a financial component, with the costs of mitigation, correction, or potential liability used as scalable factors. Litigation risk sometimes has a formal place in the risk ranking process. Reputation is also increasingly considered in environmental risk evaluations, usually part of an effort to capture risks that do not involve apparent immediate compliance or incident risks. Reputational risks are sometimes evaluated in a more granular manner, distinguishing, for example, between shareholders, customers, neighbors, public interest groups, and the public generally.

Opportunities for improvement may be considered when calibrating risk (and is a core theme of ISO 14001). There are frequently opportunities to improve performance and decrease risk even if further actions are not mandated by law. For example, even if it is legal to run a single-walled pipe carrying hazardous substances near a pond or creek, prudence may suggest identifying this as a risk, since moving the pipe or changing it to a double-walled pipe would decrease the likelihood of a release to water in the event of a leak or incident. Or one might apply Six Sigma statistical techniques to a Clean Water Act–permitted discharge to set lower internal pollutant concentration targets that account for process variability and thus decrease the likelihood of permit exceedances.

There is some controversy as to whether and how the existing degree of mitigation should be considered. One view is that mitigation should be factored in to reflect the actual current risk posed. This avoids the misimpression that a risk is higher than it is and focuses management attention on the risks for which additional mitigation is necessary. Taking mitigation into account also occasions an opportunity to evaluate the effectiveness of existing mitigation measures and identify measures to correct or improve the mitigation. Managers may get irritated if they are tagged with a major risk when they believe that they are already effectively managing it.

The contrary view is that mitigated risks might be some of the organization's most significant and that keeping them on the radar screen emphasizes the importance of maintaining that mitigation. In a well-managed company, the remaining unmitigated risks may be relatively minor, and one can arrive at an anomalous result of an EMS that focuses on relatively minor issues. It is important to keep management's eye on the ball of the major risks, even if they are currently being successfully managed, because one needs to continue to maintain or even improve the successful management of those risks. For example, explosion risks at a refinery might be effectively managed by a top-flight predictive/preventive maintenance system. However, complacency, or a trend of minor but cumulatively significant budget cuts in maintenance over time (sometimes justified by the very fact of the program's success), can reach a tipping point and cause (or contribute to) a major incident (Gold, 2019; Tita, 2019; CSB, 2007). Prudence suggests a balanced approach, where risks might be evaluated with and without taking mitigation measures into account, so that the importance of mitigation will be readily visible and its continued effectiveness can be regularly evaluated as part of the risk management process, and the mitigation measures garner continued management engagement and resources.

Each of these factors (and others not listed) has advantages and disadvantages, which recommends including several factors so that not any one dominates. How the evaluation of these factors is expressed is more a matter of organizational culture than best practice. Engineering-driven firms are often most comfortable reducing the various judgments to numbers, while other organizations work just as well with "high, medium, low" or "red, yellow, green" analyses. What is important is that the process and criteria are transparent and kept current, the results are communicated appropriately at all levels of the organization, and the balance of the system is properly designed to effectively manage the identified risks in a manner appropriate to the level of risk that they pose.

8.5 Knowing the Rules

Knowing the applicable rules is critical to successfully managing the identified environmental risks. Most environmental issues are comprehensively regulated, and a thorough understanding of legal requirements is an essential component of an effective risk assessment process, as well as of an effective system overall. This may be self-evident, but it is surprising how often major environmental liabilities have either ignorance or a misunderstanding of the law as a contributing cause. This appears to happen more frequently with smaller organizations that may not have the professional resources to keep up with the ever-growing pile of environmental regulations. It can also occur with some frequency with subjects that have not traditionally received much attention from environmental professionals, such as product regulations. Missteps can also occur because an organization does not have an accurate or comprehensive understanding of its context. For example, manufacturing or consumer products companies might not pay much attention to TSCA or REACH even if they regularly use chemicals because they are under the misimpression that the chemical regulatory requirements only apply to "chemical companies." Consumer product companies that make claims about how their products protect consumers from bacteria or mold are sometimes surprised to find that they were unlawfully marketing "pesticidal devices" regulated by U.S. EPA under the Federal Insecticide, Fungicide, and Rodenticide Act (FIFRA). Many nonindustrial retail companies were surprised by a wave of enforcement actions regarding the management of hazardous waste resulting in millions of dollars of penalties, having not focused on the fact that the rules applicable to waste generated by industrial manufacturing facilities would apply with equal force to them.

Therefore, an effective EMS must have a process for systematically identifying, keeping current with, and communicating about the legal requirements applicable to its activities, products, and services, considering the overall context within which it operates. As with the risk assessment, this should not be limited to the traditional coverage of facility environmental issues such as air emissions, water discharges, and waste management.

If this process is to be effective, it cannot rest at mere identification. Keeping the environmental professionals up to speed on legal requirements is just the beginning. The legal requirements need to be "translated" so that the information can be practically communicated to and used by the rest of the organization. Sending someone a link to a new complex regulation or a 400-page Clean Air Act Title V permit with a cover note requesting the recipient to review and comply is close to useless. Further, legal requirements should, where possible, be either explicitly or implicitly integrated into regular operational procedures. For example, a product development procedure might include a "gate" early in the process to consider the regulatory and risk status of proposed raw materials, required maintenance of pollution control devices can be entered into a computerized maintenance work management system, or business planning procedures can include triggers to ensure that potential permitting requirements are identified early in the process of planning a plant expansion. Burying this information in an environmental manual accessed primarily by environmental professionals or having a link on the organization's website to "environmental regulations" is not likely to get this information where it needs to be at the right time. Another communications issue is how the organization's environmental requirements will be communicated to third parties, including suppliers, contractors, downstream users, consumers, or the general public. These could range from raw material specifications to on-site waste management practices to instructions for use, recycling, or disposal.

An effective process will also include a forward-looking element, keeping an eye out for what requirements may be coming that could either adversely affect or benefit the organization. This information can be used for a variety of purposes, including planning ahead to adapt the organization to forthcoming requirements in a measured and planned fashion or engaging in legislative or regulatory advocacy to protect or advance the organization's interests.

It is prudent to have a broad view of the source of possible requirements and not be limited to formal legal requirements. For example, actions taken by standards bodies, trade associations, or other nongovernmental organizations may have direct and significant impacts on the organization. One need look no further than the standards developed by ISO, which can become de facto requirements as a matter of commercial necessity and in many cases have been either adopted or endorsed by governmental authorities. To provide a few other examples, Global Reporting Initiative's (GRI) criteria for sustainability reporting are widely used, and organizations that belong to the American Chemistry Council are obligated to implement Responsible Care® and the Responsible Care Management System®. ISO 14001 incorporates this concept by obligating implementing organizations to identify and conform with "other requirements," reflecting the "say what you do, do what you say" principle. As discussed in Chapter 7, the Sentencing Guidelines encourage organizations to take "industry standards" into account in the design and implementation of ethics and compliance programs.

Further, in line with the discussion above regarding delineating the overall context of the organization, it is a good idea to consider what demands or interests stakeholders may have in the organization (this too is an element of ISO 14001). These could include issues such as climate change or considering the potential contributions of the organization to the U.N.'s Sustainable Development Goals.

Another essential element is knowing and following the organization's own rules. These are not optional "second class" rules that, in a pinch, can be ignored. The organization's internal procedures, even if not legally mandated, are frequently central to the effective management of environmental risk and the prevention of incidents and noncompliance. Further, even attempting to distinguish between "legal" and "non-legal" rules is a fraught exercise, since liability issues are entwined with almost all environmental risk issues. Lax enforcement of internal procedures damages the credibility of the EMS with management and employees, suggesting that they can pick and choose what rules they want to follow. It can also damage credibility with external stakeholders: for example, it will be difficult to persuade enforcement officials that an incident was the result of isolated bad luck or a "bad apple" individual if the evidence shows that the company regularly overlooked transgressions of internal procedures.

Lastly, for all of the emphasis on rules, it is important to remember that one cannot have rules for everything. Establishing a strong cultural foundation of "doing the right thing" even in the absence of a specific rule can go a long way toward decreasing the likelihood of misconduct or incidents.

8.6 Objectives, Programs, and Procedures: Figuring Out What to Do

Having established the basic policy direction of the organization and identified the environmental risks and rules, the next step (assuming, which is not always the case, that one is developing an EMS in a linear fashion) is to translate this into measurable objectives that describe what the organization intends to achieve.

Compliance is a minimum target. There is a view that compliance should be assumed and does not merit a discrete objective. While that is an admirable thought, it is difficult to square with a senior management commitment to comply, the extent to which compliance is an integral component of environmental performance, the drastic consequences of noncompliance, and the explicit expectations of governmental authorities regarding compliance programs. One might also consider the internal impact of excluding compliance from environmental objectives while at the same time claiming that compliance is a core value.

Omitting key issues such as compliance from objectives can sometimes be a consequence of taking mitigation measures into account and focusing explicit objectives solely on those issues for which additional mitigation is deemed necessary. While this makes sense from an engineering or project management perspective, it can have adverse risk management consequences. For example, a successful organization with an excellent compliance record might find itself with only a few minor objectives associated with some residual low-risk issues. If queried about the absence of any major issues, the response might be "we have already taken care of those." That raises the question of how the performance on the key issues will be sustained or improved if they no longer appear on management's radar screen as targets or objectives. It is possible to design EMS to distinguish between identifying and completing remaining mitigation projects, regardless of their overall importance, and retaining the necessary emphasis on maintaining or improving performance on high-priority/risk matters. Whatever approach is taken, it is important that maintaining performance on high-risk issues is not taken for granted and somehow becomes invisible in the system, at which point one is left with the asymmetrical situation of an EMS paying more explicit attention to lower-risk matters. DOJ has identified this as a factor it will look at in evaluating systems.

Objectives should also reflect organizational commitments other than just those related to compliance and related environmental risks, such as those related to sustainability or corporate responsibility. For example, objectives might include reducing greenhouse gas emissions, increasing the use of recycled or reused materials, improving environmental or working conditions in its value chain, eliminating certain chemical substances from products, enhancing the efficiency of certain product lines, etc. Objectives might also go "beyond compliance" in an effort to improve environmental and compliance performance. For example, an organization might set a target to reduce all spills of any quantity, regardless of whether they are reportable to regulatory authorities or violate the law, in an effort not only to protect the environment but also to reduce the likelihood of larger spills. These objectives can be expressed into measurable key performance indicators (KPIs), or some similar business measure, that can trickle down from corporate goals all the way to individual KPIs. In this way, measurable environmental objectives can be integrated into and tracked with other business goals.

In order to consistently meet objectives, one must identify who is responsible or accountable for success, how the objective will be achieved, what the milestones and schedules will be, and what resources must and will be applied to the effort. (In ISO 14001 parlance, these are called programs.) In some cases, achieving the objective will be a defined project management task with a fixed beginning, middle, and end, while in other cases it will effectively be a permanent project with no end date. For example, compliance with most regulatory programs will have constant programs (e.g., waste management, chemical management), as will programs aimed at emergency management and addressing other unplanned events. The point here is that it is not credible to issue diktat regarding objectives without having defined programs on how those objectives will be achieved and who is responsible.

In the "old days," the answer to the question of "who is responsible for environmental compliance" was answered by pointing to the facility environmental professional. Times have improved since then, with the recognition that the environmental professional is in a staff position with no authority or control over the individuals or operations that generate the environmental risk in the first instance. Further, unless one wants to expend the resources to have a "cop on every corner," the environmental staff function simply does not have the resources to ensure compliance or the successful management of all risks. The recognition that "everyone is responsible for compliance" calls for the distribution of environmental responsibilities throughout the organization, placing responsibilities on the functions and management responsible for creating the compliance obligations in the first instance. This cannot be done by simple declarations; otherwise, "everyone is responsible" translates to "no one is responsible." If the product design function or the real estate development group are going to be given environmental responsibilities, they need the procedures (or, using ISO 14001 terminology, operational controls), training, assistance, etc. to fulfill those responsibilities. Management engagement is also a central element of this process: the "everyone is responsible" mantra does not work if management does not demonstrate by their daily engagement that this applies to them as well.

Some organizations manage this procedural challenge by creating standalone environmental procedures, sometimes assembled (physically or electronically) into an environmental manual. This has the benefit of having all the environmental requirements in one place for purposes of drafting, control of over the documents, and general access. For those focused on demonstrating the system to third parties, whether they be governmental authorities or ISO 14001 registrars, this option may be attractive. The shortcoming of this approach is that the necessary environmental actions/procedures might not be well integrated into the business or operational process itself. Users may be obligated to pay attention to two sets of procedures and figure out how the two are to mesh together. Creating parallel procedures might unnecessarily complicate matters where the insertion of just a few sentences or paragraphs in existing procedures might do the trick.

As is frequently the case, many organizations take a middle approach, creating free-standing environmental procedures where appropriate and inserting environmental requirements in existing operational procedures where that will work best. Another approach is to insert links or electronic gates into existing procedures to either require an environmental review before the process can continue or direct the process to an environmental professional who must review the status of the process and give it the green light before it may proceed. It is also common to insert a variety of environmental equipment and inspection requirements into existing work or maintenance scheduling programs (e.g., the popular Maximo program). Some organizations apply lessons learned from well-accepted business systems, such as lean manufacturing, to inform the development of environmental controls. Procedures or controls do not have to be complex or long. Waste management procedures might be conveyed through simple pictures or diagrams on a poster, and daily reminders to ask about common environmental issues can be added to documents used to guide morning-shift meetings or pre-job briefings. The concept is straightforward: when people show up for work, they should have the instructions and tools at hand to do the job correctly.

Addressing third parties, such as contractors, suppliers, vendors, etc., is a sensitive subject in developing and executing procedures and operational controls. In the current context of legal requirements, litigation risks, and social expectations, it is difficult, if not impossible, for organizations to have a completely "hands off" attitude with respect to their value chain. On the other hand, if the direction becomes too intrusive, the organization may expand its liability risk if it exposes itself to "piercing the corporate veil" or find that contractors had been transformed into employees. However, the failure to adequately coordinate and control the activities of contractors and vendors can have significant consequences (CSB, 2016). Virtually all comprehensive EMS include robust programs for addressing the environmental risks associated with third parties, but calibrate them so as not to unintentionally increase legal risks.

Management of change is a theme running through all programs, procedures, and controls. Any number of environmental incidents can be traced to inadequate change management, whether that be using outdated procedures, having employees change jobs but not receive training on their new responsibilities, purchasing a new operation or business without adequate diligence, or changing operations or equipment without adequate review of the potential consequences of the change. An effective system will include processes and gates for capturing, evaluating the consequence of change, and taking the necessary measures to prevent or mitigate adverse consequences. Change management should also consider changes that might occur over time through a series of what might appear to be minor events. For example, incremental changes and reorganizations of the work, whether planned or occurring for other reasons, such as demographics, might eventually, and with hindsight, reveal such significant weakening of functions such as maintenance to have contributed to a major adverse event.

Management of change procedures should also be considered for significant changes in legal requirements. For example, some organizations have defined processes for evaluating operational changes necessitated by new permits or regulations that will be initiated long before these requirements are finalized, giving the organization a reasonable opportunity not only to prepare for the new requirements but also to participate in advocacy at the early stages of the development of the legal requirements. This may avoid, for example, situations where a final permit contains a requirement with which the facility cannot comply, but the organization never commented or raised the issue during the permitting process and is thus in no legal position to challenge the permit.

As discussed in Chapter 7, organizations should recognize the limits of procedures. There cannot be a procedure for every eventuality, and it is typically dysfunctional to attempt to create procedures with layer upon layer of detail that attempt to be all-inclusive. That is one of the benefits of having the EMS be part of or aligned with the underlying ethics and compliance system, which insists on a culture of ethical and compliant conduct (i.e., do the right thing) even in the absence of a specific rule.

8.7 Measuring Performance, Auditing, and Corrective/Preventive Action

"What gets measured gets done" is a fundamental management principle. Data and metrics drive organizational and management attention, performance, and action. If environmental performance is not frequently measured, it will likely get less consistent management attention (until some big incident occurs) and signal that environmental issues are relatively less important. When KPIs are all about production, quality, delivery, and finance, and there is nothing about environmental, that can take environmental off the radar screen.

An effective EMS regularly measures the organization's environmental performance. If the EMS is integrated with the organization's business or operational systems, environmental information can be folded into the organization's regularly reviewed metrics. At the management engagement level, this at a minimum should include performance against the defined objectives and incidents. This is a significant departure from the "report by exception" of the traditional approach, in which significant environmental information would go up the management chain only when something went wrong. Environmental information can be discussed on an almost real-time basis along with other business critical information at, for example, daily, weekly, or monthly meetings.

This feature of "operationalizing" environmental management is important to success. Direct and regular management engagement, driven by everyday processes, not special environmental meetings convened by environmental professionals, weaves environmental issues into the daily narrative of the organization. In this model, the primary and initial responsibility for generating, evaluating, and acting on the environmental data typically lies with operational management, with environmental professionals playing an important, but still staff support, role. This bears similarities to economic and financial issues: operational management are ultimately responsible for their numbers. In addition, the system will include such monitoring and measurement as is necessary for legal or technical purposes (e.g., permit monitoring and waste characterization).

A cautionary note: the corollary to "what gets measured gets done" should not be "what is difficult to measure is not important." The importance of an environmental issue or the extent of an environmental risk should not be based on its susceptibility to easy quantitative measurement. A purely numbers-driven approach can lead to significant issues being overlooked or mismanaged. Similarly, organizations should recognize that complex environmental risk issues may be interpreted differently in the public square than in their own laboratories or boardrooms. The reputation and financial security of companies can founder on the shoals of public opinion and jury verdicts even as their scientists, and perhaps even those of the regulators, have concluded after long and sober analysis that a product does not pose an unreasonable risk.

Discrete from the ongoing performance measurement is auditing to verify compliance and conformance to the requirements of the system. Compliance auditing has long been with us, and the core elements of an effective auditing program are well known and do not need too much discussion here. This is not to diminish its importance to the organization and the effectiveness and credibility of the EMS, particularly to governmental authorities. A weak compliance auditing program will be exhibit number one supporting the government's conclusion of an inadequate compliance program.

The scope of the auditing program should be consistent with the scope of the organization's overall environmental risk management system. Thus, it should not be limited to the traditional scope of auditing facility manufacturing or assembly activities. It should cast a broad net, evaluating compliance with applicable product regulations, transportation and distribution, etc. To the extent that the organization has made extra-legal commitments on issues such as reducing greenhouse gas emissions, cutting water consumption, etc., those should be within the scope of the audit program as well.

Audit programs should also consider the information from related compliance-checking activities, such as less formal inspections, facility self-audits, etc. First, that provides more information gathered on a more frequent basis for purposes of evaluating the organization, trend analyses, etc. It also can help evaluate the effectiveness of regular checks and controls: if a second- or third-party audit identifies many new findings, one of the first questions asked should be why the organization (or facility) had not already detected, prevented, and corrected these issues.

As discussed in Chapter 7, the design and execution of the compliance audit program should consider the availability of federal and state audit privilege and voluntary disclosure programs. EPA provides an opportunity for either no or significantly reduced gravity-based penalties for violations that are detected systematically through the operation of an auditing or compliance program, and promptly (generally within 21 days of discovery) and voluntarily reported to EPA (i.e., if disclosure is mandatory, it generally will not qualify). EPA has special rules that provide incentives for new owners of businesses to voluntarily disclose pre-closing noncompliance that occurred during the prior owner's watch that should be taken into account as part of mergers and acquisitions, with respect to both pre-closing due diligence and the scheduling of any audits or compliance reviews immediately after closing. *Interim Approach to Applying the Audit Policy to New Owners*, 73 Fed. Reg. 44991 (Aug. 1, 2008); *New Owner Clean Air Act Audit Program for Oil and Natural Gas Exploration and Production Facilities* (March 29, 2019). Many states have their own audit and disclosure programs that, while generally similar to the EPA policies, may have their own quirks. For example, under the Texas Environmental, Health and Safety Audit Privilege Act, organizations must give the state prior written notice of the planned audit to qualify for its benefits. Therefore, it is a good practice to consider these federal and state policies when designing an audit program or planning a specific audit, rather than thinking about it for the first time when contemplating draft adverse audit findings.

Even if all the criteria of these policies are not met, it may be possible to qualify for a reduced penalty, if not a 100% reduction. Separate from audit policies, EPA has subject-matter-specific "penalty policies" that give the Agency discretion to significantly reduce penalties from the statutory maximums, considering factors such as the nature and extent of the violation (i.e., gravity), prompt (typically within 30 days of discovery) voluntary disclosure, and degree of cooperation. It is not uncommon for organizations making voluntary disclosures to argue that they met the criteria for a 100% reduction of penalties under the relevant audit policy, but, even if those criteria were not fully satisfied, any penalty should be significantly reduced under the applicable penalty policy. The audit and enforcement policies are voluntary enforcement discretion policies, not mandatory reporting requirements (i.e., it is not a violation of law to not take advantage of these policies). Organizations may and do exercise their judgment as to whether to take advantage of them.

The organization must audit its EMS to verify that it is well designed, is fit for purpose, and is being effectively implemented. This is a discrete objective from performance measurement or compliance auditing. The systems audit is aimed at verifying that the system that produces the environmental performance can be trusted. There are two basic objectives: (1) determining if the EMS has been properly designed and (2) verifying that it has been effectively implemented as designed. The first inquiry starts by comparing the system to whatever criteria have been chosen by the organization, which could be the Sentencing Guidelines, DOJ guidance, ISO 14001, or any number of other criteria that could be implemented in countless combinations. One should not be wedded to such criteria at the expense of reality. For example, a system might be brilliantly designed consistent with well-recognized criteria, but elements of that design might be ill-suited for the culture of that organization. Information learned in practice in the field is a good feedback loop for improving (and frequently simplifying) the design of the EMS. Accordingly, one could conclude that the system has not, in some respects, been well designed even if it might check all the ISO 14001 or Sentencing Guidelines boxes.

The second inquiry is verifying the effective implementation of the EMS. This is not a paperwork or desktop review. Implementation is verified by checking that the actions required by the system are being undertaken and the intended performance is being achieved. The scope of this exercise is as broad as the system itself. This can extend the review beyond compliance issues into other matters included in the environmental risk management system, such as sustainability issues and programs intended to achieve targets not mandated by regulations (e.g., reductions in greenhouse gas emissions or reductions in the use of water). If the organization regularly publishes a sustainability report, that would (and should) also be fair game for an EMS audit.

Objective evidence must be gathered confirming that the EMS is being implemented as planned and that the EMS is achieving the desired result. This review is not limited to reading procedures and records: it should include interviews, observations, and other methods to directly verify implementation. For example, the review of a training program should not be limited to reviewing training matrixes and training records that demonstrate that members indeed attended the required training sessions. A selection of individuals should also be interviewed to verify that they know what they are doing, are competent, and are aware of the issues relevant to them. In other words, is the training effective? The review of procedures should include, on a selective basis, walking through each step of a procedure with interviews, facility observations, record reviews, etc., to verify how everything actually works. If a preventive/corrective action data base is being reviewed, do not assume the entries reflect reality. Spot check some of those entries to verify that what has been recorded as done was done, or that it is realistic that a complex corrective action plan is likely to be completed as scheduled.

The focus is on conforming to the requirements of the EMS as much as it is on compliance. The fact that the deviation from a procedure or the failure to meet a target may not violate the law does not make it any less a finding or relegate it to a low-priority "observation." Indeed, detecting and correcting shortcomings in the system (e.g., training, risk assessment, contractor management) before they have ripened into a serious incident or major noncompliance is a central function of an effective systems auditing program. Assessing performance also is a core element of an EMS audit. The effectiveness of an EMS can only be judged if the review includes an evaluation of the performance that the EMS is supposed to produce. There is a serious disconnect if an auditing program concludes that the EMS is properly designed and effectively implemented if the organization is not hitting its environmental targets or has extensive compliance audit findings.

An effective EMS auditing program can also significantly contribute to the ongoing improvement of the system, not just detect problems. Continual improvement is a central theme of ISO 14001, is integral to business systems such as lean manufacturing and operational excellence, and has been highlighted as well by DOJ:

> One hallmark of an effective compliance program is its capacity to improve and evolve. The actual implementation of controls in practice will necessarily reveal areas of risk and potential adjustment. A company's business changes over time, as do the environments in which it operates, the nature of its customers, the laws that govern its actions,

and the applicable industry standards. Accordingly, prosecutors should consider whether the company has engaged in meaningful efforts to review its compliance program and ensure that it is not stale. Some companies survey employees to gauge the compliance culture and evaluate the strength of controls, and/or conduct periodic audits to ensure that controls are functioning well, though the nature and frequency of evaluations may depend on the company's size and complexity.

Prosecutors may reward efforts to promote improvement and sustainability. In evaluating whether a particular compliance program works in practice, prosecutors should consider "revisions to corporate compliance programs in light of lessons learned."

(DOJ 2019, 14)

In this way, the outputs of the EMS auditing process, as with other information generated by the EMS, contribute to effective management engagement and continual improvement.

One of the ways performance and auditing information can be used is to evaluate it for trends, both positive and negative. Organizations are taking advantage of "big data" analysis tools to extract useful trends and lessons out of heretofore unexploited information. This is yet another way that EMS can take advantage of existing business systems that may already be improving overall organizational performance using these tools.

Organizations must have procedures for addressing detected violations or nonconformities. In colloquial terms, an organization must look, find, and then fix what it finds. There are several elements to consider. The immediate situation must be corrected. The consequences should be mitigated. The cause of the problem should be determined, with the intensity of the root cause analysis tailored to the nature and extent of the issue. Based on this information, actions should be taken to prevent the recurrence of the problem. A comprehensive corrective and preventive measures procedure may allow for different levels of response depending on the situation. The formality of each of these steps may vary depending on the nature and extent of the issue at hand. Sometimes these steps might be accomplished very quickly, while in other situations enterprise-wide and capital-intensive corrective and preventive measures may ensue. Some organizations draw these procedures narrowly, addressing only the results of formal audits. This has the benefit of clarity and focus, though it can leave out of the corrective/preventive measures procedure information that is equally or maybe even more important than that generated by audits and thus provide an incorrect picture of the true status of the organization. For example, "big data" trend analyses may surface issues that merit enterprise-wide preventive/corrective action well before any significant incident or noncompliance occurs. Procedures for root cause analyses are sometimes embedded in the corrective/preventive action procedure, while other organizations will separate this element out. Again, it is common for a root cause analysis procedure to have different levels of intensity and engagement depending on the severity of the issue at hand.

Corrective and preventive action may, depending on the outcome of the root cause analysis, include changes to the underlying system itself. Simply immediately correcting the visible problem may miss and potentially even conceal a more serious systemic issue that has not yet materialized in a serious incident. The goal is to understand why an incident occurred, not just how it occurred. An effective root cause analysis will not stop at the first obvious cause and keep asking "why" until the cause of an incident is truly understood. Simply concluding that an employee did not follow procedures without understanding why that happened leaves unaddressed a range of issues, including the adequacy of training, oversight, procedures, resources, etc. In this manner, preventive action is aimed not simply at preventing the recurrence of noncompliant events that have occurred, but also at preventing them from occurring in the first place. It is a good day when a systems audit detects and triggers the correction of a systems deficiency before it has ever ripened into even a small incident. The fact that a systems deficiency may not itself be a violation of a regulatory requirement does not make it any less important to correct. Placing a lower priority on correcting "systems observations" if they are not tied to a specific instance of regulatory noncompliance is a mistake.

An organization's auditing and corrective/preventive action programs should be supplemented by a procedure for investigating detected or potential misconduct. Effective investigation of misconduct is very important to demonstrating the credibility of the overall program to governmental authorities. For example:

> Another hallmark of a compliance program that is working effectively is the existence of a well-functioning and appropriately funded mechanism for the timely and thorough investigations of any allegations or suspicions of misconduct by the company, its employees, or agents. An effective investigations structure will also have an established means of documenting the company's response, including any disciplinary or remediation measures taken.
>
> **(DOJ 2019, 15)**

It is possible to integrate such a process into an organization's corrective/preventive action programs. However, it is likely that most organizations will conduct discrete investigations of environmental misconduct using the procedures available under the general ethics and compliance program. These are typically conducted under the supervision of the general counsel's office (or outside counsel).

This raises another issue that should be considered when designing audit and corrective action procedures: privilege and confidentiality. The traditional approach is to conduct all audits under the direction of in-house (or outside) counsel and closely hold the results on a "need to know" basis to preserve attorney-client privilege and thus best protect the organization from liability. Many organizations now find this to be a burdensome and often counterproductive process that impedes the flow of necessary information and decreases the effectiveness of the EMS. In response, some organizations have eliminated privilege from their auditing processes, opening the flow of information (though typically keeping lawyers involved in the process to offer legal interpretations, assist with conclusions as to whether a situation constituted a violation, etc.). Other organizations have tried to find a middle ground. For example, some will conduct the audit under privilege, but after prompt legal review of draft findings "declassify" and finalize the report except for sensitive issues that might be carved out for further review under privilege. A variation on that theme is initiating the audit without the privilege umbrella, but immediately segregating potentially sensitive issues upon discovery for separate investigation under privilege.

Privilege is an issue for EMS audits as well as compliance audits. A "systems finding" can potentially be more sensitive than a straightforward noncompliance finding. For example, a finding of inadequate management engagement or insufficient funding could take one down a trail toward potential criminal liability if paired with noncompliance findings. Organizations should also be aware of the potential value of privilege during root cause analyses, during which sensitive issues may bubble up. This is not to suggest a default rule that all root cause analyses should be conducted under privilege. Rather, it is something to consciously consider so that one does not run into situations where lawyers are asked, typically inappropriately, to provide an after-the-fact "privilege blessing" for a completed audit or root cause analysis that someone just figured out included some very sensitive information.

8.8 Training, Awareness, and Communication

The EMS will not work if no one knows about it. Establishing reasonable job requirements (e.g., based on education and experience), combined with effective training and awareness, are critical to the success of an EMS. The organization should have a defined procedure for establishing competency requirements and identifying training needs, including who needs to receive the training, as well as the content and schedule for that training. This may extend beyond the organization's own members and include, with varying degrees of intensity and detail, contractors, suppliers, and others (e.g., visitors). The goal is securing the necessary degree of competence and awareness so that the correct people are at the job for which they are qualified and know what to do. Thus, the effectiveness of a training program cannot be measured solely with training records showing that individuals showed up at (or signed up for) the correct classes. An EMS audit should interrogate the outcomes to verify that people know what they are doing, not simply check the box that the training records are complete.

It is important that this competence and awareness goes beyond just technical issues and encompasses the EMS and the organization's employees' role in it. This is necessary if organizations are going to move from a "the environmental professional oversees environmental" model to a more dispersed approach in which a broader group of operational management and employees have important responsibilities. It will not work to simply declare that operational management is primarily responsible for environmental issues without providing them the tools, including effective training and education, to make this practically possible. Part of this process, in addition to training, is regular communications about environmental issues, including about performance, audit results, expectations, developing requirements, etc. Organizations frequently have electronic "dashboards" or analogue bulletin boards on which a wide range of business performance information is posted and updated. There is no reason why environmental and sustainability information cannot be added. In other words, include relevant environmental information in the normal flow of business information.

As with the emphasis on internal investigations, governmental criteria for good communications procedures go beyond what is called for by standards such as ISO 14001 and direct organizations to implement internal reporting procedures (e.g., "hotlines" or "helplines") that allow employees to report compliance concerns or ask questions without fear of retaliation. In many organizations, this criterion is addressed outside of the EMS as such and within the overall ethics and compliance program discussed in Chapter 7.

Effective environmental risk management systems increasingly have an external communications strategy as well. With environmental risk footprints now determined not solely based on an organization's internal priorities and analysis of applicable law, but also on inputs from a range of interested stakeholders and the public in general, it is naïve to attempt to effectively manage environmental risks without a planned and active communications strategy. The advent of lightning-fast social media, in which an apparently minor incident or even incorrect information can quickly blossom into a torrent of public attention or indignation, makes it prudent to have a systemic approach to try to stay ahead of the curve.

It is common now for organizations to communicate externally about their environmental issues, with links to information or reports on their website perhaps the most common. Sometimes organizations' websites direct users to detailed sustainability reports (often in the format established by the GRI), and social media outlets have been used to distribute environmental information. The Sustainability Accounting Standards Board (SASB) published sustainability accounting standards in 2018 (www.sasb.org). This information supplements the rather large volume of environmental information that under U.S. law must be made publicly available anyway, including environmental disclosures under the environmental statutes (e.g., the Emergency Planning and Community-Right-To-Know requirements) as well as disclosures about environmental issues in securities filings (e.g., to the Securities and Exchange Commission) (SASB 2017a). There has been steadily increasing pressure, first from public interest groups and shareholders, and now from other financial stakeholders (e.g., insurance companies), for organizations to disclose more information about the potential impact of climate change on their operations and finances. The alleged failure to make accurate or complete disclosures regarding environmental risks can become fodder for litigation by disappointed investors or irritated regulators (SASB 2017b; *In re BP P.L.C. Sec. Litig.*, 922 F. Supp. 2d 600 (S.D. Tex. 2013). Court held that plaintiffs' allegations that statements in annual reports and sustainability reports misrepresented BP's environmental risk management practices were sufficient to withstand a motion to dismiss a securities fraud action arising from the Deepwater Horizon incident; the courts later concluded that the plaintiffs could not be certified as a class, *Ludlow v. BP, PLC*, No. 14-20420 (5th Cir. September 8, 2015)).

Given these various avenues for external communication about environmental issues, and the speed with which information can spread, it is prudent, where possible, to coordinate these activities in a disciplined manner to ensure that the information is accurate, consistent, and can be credibly supported with on-the-ground performance. One should avoid conflating sustainability with talking or reporting about sustainability. Many public measures or rankings of sustainability rely heavily on the information contained in sustainability reports or website links, and thus there can be pressure on corporate communications or environmental departments (or both) to deliver a persuasive sustainability report or communications package. However, these efforts should not "get too far over their skis" and suggest levels of performance or commitment that cannot be supported on the ground. Starting with a strong system that consistently delivers results and then talking about it is generally a better strategy than trying to do it the other way around.

8.9 Documents and Records

The core elements of the EMS should be documented so that the users can use and refer to it. While not every single control necessarily requires a discrete procedure, a balance should be achieved so that the organization is not relying on "oral tradition" to effectively identify and manage its environmental risks. Having a reasonably well-documented system also facilitates demonstrating its effectiveness to third parties, including governmental authorities.

Interestingly, perhaps in response to long-standing complaints about ISO management systems standards and the related registration process being too focused on paperwork, the 2015 revisions to ISO 14001 decreased the documentation requirements and provided more flexibility. "Document control" adds further discipline by requiring that the correct versions of the necessary procedures be available to the right people at the right time at the right place so that they will do the right thing. Inaccurate or out-of-date procedures, or hand-written notes on yellow stickies posted on control boards (i.e., uncontrolled documents), do not enhance effective risk management.

The organization must also have a record-keeping system that meets regulatory requirements and is also sufficient to demonstrate that the system itself is being implemented. Both the document control and records management procedures might be common procedures shared by several functions in the organization, including production and quality.

8.10 Conclusion

It is possible to, and many organizations still do, manage environmental risks on an episodic, ad hoc basis. This might still work for organizations with very little exposure to any environmental risks, and perhaps for organizations who have thus far been lucky. However, for the vast majority of organizations, it makes common, financial and legal sense to identify and manage environmental risks through a disciplined and comprehensive system that, where possible, is consistent and integrated with other core business systems. In principle, such systems, typically based on the simple plan-do-check-act model, are not that complicated. The challenge lies less in the design than in the execution, which requires the direct and continuous engagement of the organization's management and members.

References

American Chemistry Council. "Responsible Care—Management System and Certification." Accessed November 29, 2019. https://responsiblecare.americanchemistry.com/Management-System-and-Certification.

Business Roundtable. "Statement on the Purpose of a Corporation." August 19, 2019. Accessed November 29, 2019. https://opportunity.businessroundtable.org/ourcommitment/.

Department of Defense (DOD). *Departmental use of Environmental Management Systems*, October 24, 2016. https://www.denix.osd.mil/ems/home.

Department of Justice (DOJ). *Evaluation of Corporate Compliance Programs—Guidance Document, Updated April 2019*, 2019. https://www.justice.gov/criminal-fraud/p age/file/937501/download.

Environmental Protection Agency (EPA). "Environmental Management Systems." Accessed November 23, 2019. www.epa.gov/ems.

EPA. "EPA's Audit Policy." https://www.epa.gov/compliance/epas-audit-policy.

EPA. "New Owner Clean Air Act Audit Program for Oil and Natural Gas Exploration and Production Facilities." 2019. https://www.epa.gov/enforcement/new-owne r-clean-air-act-audit-program-oil-and-natural-gas-exploration-and-production.

Global Reporting Initiative. Accessed November 29, 2019. www.globalreporting.org.

Gold, R., R. Smith, and K. Blunt. "PG&E: Wired to Fail." *Wall Street Journal*, December 28, 2019. Accessed January 6, 2020. https://www.wsj.com/articles/pg-e-wired-to-fail-11577509261?mod=searchresults&page=1&pos=14.

International Organization for Standardization (ISO). "ISO 14001: 2015, Environmental Management Systems—Specification with Guidance for Use." 2015.

ISO. "ISO 14004: 2016, Environmental Management Systems—General Guidelines on Implementation." 2016.

ISO. "ISO Survey 2018." 2018. https://www.iso.org/the-iso-survey.html.

ISO. "ISO 14005: 2019, Environmental Management Systems—Guidelines for a Flexible Approach to Phased Implementation." 2019.

O'Neil, C. *Weapons of Math Destruction: How Big Data Increases Inequality and Threatens Democracy.* New York: Crown, 2016.

Ramani, V., and H. Saltman. "Running the Risk: How Corporate Boards Can Oversee Environmental, Social and Governance (ESG) Issues." *Ceres*, November, 2019.

Sustainable Accounting Standards Board (SASB). The State of Disclosure – An Analysis of the Effectiveness of Sustainability Disclosure in SEC Filings. 2017(a). file:///C:/Users/bellc/Documents/NEW/Research/StateofDisclosure -Report-web112717-1.pdf

Sustainable Accounting Standards Board (SASB). Legal Roundtable on Emerging Issues Related to Sustainability Disclosure, 2017(b). Accessed May 10, 2020. https://www.sasb.org/wp-content/uploads/2019/08/LegalRoundtable-Paper-11132017.pdf

Tita, B., and K. Maher. "U.S. Steel, the Company that Built America Faces Its Age." *Wall Street Journal*, December 15, 2019. Accessed January 6, 2020. https://ww w.wsj.com/articles/u-s-steel-the-company-that-built-america-faces-its-age-1 1576443004?mod=searchresults&page=3&pos=2.

United Nations. "Sustainable Development Goals." Accessed November 23, 2019. https ://www.un.org/sustainabledevelopment/sustainable-development-goals/.

U.S. Chemical Safety and Hazard Investigation Board (CSB). *Investigation Report, Refinery Explosion and Fire, BP Texas City*, March 20, 2007. Accessed January 6, 2020. "https://www.csb.gov/bp-america-refinery-explosion/" https://www .csb.gov/bp-america-refinery-explosion/.

U.S. Chemical Safety and Hazard Investigation Board (CSB). Investigation Report, Drilling Rig Explosion and Fire at the Macondo Well. April 17, 2016.

U.S. Chemical Safety and Hazard Investigation Board (CSB). CSB Best Practice Guidance for Corporate Boards of Directors and Executives in the Offshore Oil and Gas Industry for Major Accident Prevention. April 29, 2020.

9

Practical Methods to Solve Environmental Problems and to Reduce Risks

John Voorhees

9.1 How Alternative Dispute Resolution Works

As explained in another chapter, environmental litigation is expensive and deleterious to business interests and can be and should be avoided whenever possible. In sophisticated and complex environmental cases, lawyers', consultants', and experts' fees mount quickly. In addition to the expenses involved, backlogs and other delays frequently occur so that resolution of cases takes many months or, in most cases, even years. Litigation is always disruptive to business interests, and outcomes are never certain. Settlement in most civil cases occurs after discovery is completed and a trial date is quickly approaching. By that time, the parties have become so polarized that collaborative outcomes are unlikely and relationships, to the extent that they exist, are strained, if not broken forever.

In environmental cases, there are additional compelling reasons to avoid protracted litigation. The judiciary, as good as it is, lacks technical training and oftentimes has difficulty adjudicating the complex scientific evidence that is frequently the main subject matter of environmental disputes. The Federal Judicial Center recognized this problem and published the *Reference Manual on Scientific Evidence* to assist judges in managing expert evidence, primarily in cases involving issues of science and technology. When the government brings a business to trial court for an environmental enforcement or civil action, it is rare that the business prevails. Most victories for businesses, and they are rare, occur in trials or in the courts of appeals after lengthy and contentious trials and expensive appellate proceedings. Many business leaders have understandably become intolerant of these uncertainties, in part because litigation is controlled by judges and lawyers. Others are frustrated with the transactional costs and delays in solving problems and disputes. While courts decide legal questions, they do not and cannot resolve all disputes, nor can they assure in all cases that parties are justly compensated for cognizable legal injuries. A successful defense, though rare, can deplete corporate financial and human resources and leave managers jaded by the experience just as an unsuccessful defense can wreak havoc on business operations.

Litigation often polarizes the parties as each side is encouraged to argue its position in a self-serving manner, emphasizing the degree of injury or its lack thereof (Dauer 1994). In these circumstances, collaborative outcomes are less likely. Litigation also results in widespread public scrutiny, which can have serious adverse financial and public relations consequences. The Monsanto "Round Up" litigation of 2019 is a classic example of adverse consequences. These disruptions divert management's attention from corporate goals and objectives. For all these reasons, lawyers and business people are constantly seeking new and more effective ways to resolve disputes, cost effectively and with finality.

In 1990 Congress passed the Administrative Dispute Resolution Act, requiring each federal agency to adopt a policy that addresses the use of alternative means of dispute resolution and case management. On October 23, 1990, President George H.W. Bush signed an executive order requiring the use of alternative dispute resolution (ADR) as a tool to enhance the negotiating process and to avoid the disadvantages to all parties of protracted litigation. The year before the Act was passed, 220,000 civil suits were filed in U.S. courts, and the federal government was a party in 55,000 of these cases. On February 5, 1996, President Clinton signed an executive order directing federal agencies to implement several strategies, including the use of ADR, to resolve civil claims by or against the government. The authorization of ADR was detailed in the Alternative Dispute Resolution Act of 1998, and in 2000, the Equal Employment Opportunity Commission (EEOC) required all federal agencies to make an ADR program available during the pre-complaint and formal complaint stages. At present, all agencies are still required to have an ADR program, and most use mediation as the preferred ADR process.

ADR is becoming a preferred method to assist parties in resolving disputes before litigation is filed quickly and efficiently without the need for court intervention. In Colorado, thousands of corporations have adopted the ADR policy statement that provides:

> In the event of a business dispute between our company and another company which has made or will make a similar statement, we are prepared to explore with that party resolution of the dispute through negotiation or ADR techniques before pursuing full-scale litigation. If either party believes that the dispute is not suitable for ADR techniques, or if such techniques do not produce results satisfactory to the disputants, either party may proceed with litigation.

(Center for Public Resources Policy Statement)

Different forms of ADR include early neutral evaluation, arbitration, mini and summary jury trials, and mediation. Early neutral evaluation involves a judge, a magistrate, or an outside person who provides a nonbinding assessment of the case. Arbitration involves an arbitrator, or a panel of arbitrators, who listens to presentations by both sides and then renders a judgment based on the law and facts. A mini trial is an abbreviated presentation of the facts to a mediator who may be joined by high-level principals from each side. The process is known as a summary jury trial when the presentation is made before a mock jury that renders a nonbinding verdict (Schiffer and Juni 1996).

As many environmental cases have been resolved using a mediator, this is the primary ADR technique discussed in this chapter. Mediation uses a third party, with no decision-making authority, to assist disputants in reaching a voluntary negotiated settlement. Mediators are usually former judges available through mediation services, local lawyers who mediate from time to time as an adjunct to their practices, and magistrates and presiding judges (Schotland 1995). Mediation is a voluntary and confidential settlement procedure in which either party can back out at any time. Both sides have to voluntarily agree upon a mediator who assists the parties in reaching consensus without incurring the runaway costs of litigating the case to a verdict. Early resolution usually achieves the best financial results for all those involved. For businesses, this means scarce resources will be preserved and employees' time will be spent on more useful activities. There is also less likelihood that the corporate image will be tarnished in the marketplace and global community. The three most common advantages of mediation are that it helps bridge gaps between huge demands and low offers; it affords each party an opportunity to have a realistic appraisal of the strengths and weaknesses of its case by a neutral, detached, and experienced third party; and it defuses personality conflicts that so often arise in the context of litigation and interfere with settlement prospects (Schotland 1995).

For all these reasons, mediation works well in resolving civil disputes either before or after a case is filed. In the Florida state court system, it has been reported that between 60% and 70% of all civil cases settle during or shortly after mediation (Schotland 1995). For Chevron, ADR-based mediation of one dispute cost $25,000, whereas going to court would have cost as much as $2.5 million (Carver and Vondra 1994). Toyota set up a Reversal Arbitration Board to resolve disputes between the company and its dealers concerning the allocation of its cars and sales credits to the dealers. This ADR procedure reduced the number of cases from 178 in 1985 to 3 in 1992 (Carver and Vondra 1994). Using ADR techniques, AT&T Global Information Systems reduced the number of filed lawsuits from 263 in March of 1984 to 28 in November of 1993 (Carver and Vondra 1994).

The Environmental Protection Agency (EPA) has wisely been advocating the use of ADR to resolve environmental disputes before litigation is filed. In May 1995 the EPA reported that it employed this technique to assist in the resolution of over 50 enforcement-related disputes. These cases range from two-party Clean Water Act actions to Superfund cases involving up to 1,200 parties. Participants in an ADR pilot for Superfund cases reported the following benefits: "constructive working relationships were developed; obstacles to agreement and the reasons therefor were quickly identified; costs of preparing a case for U.S. Department of Justice referral were eliminated; and ongoing relationships were preserved" (EPA 1991). The EPA insists that ADR techniques, including mediation, provide other benefits that will minimize transactional costs in resolving disputes. In 1999, the EPA created its Conflict Prevention and Resolution Center (CPRC) to expand and improve upon its utilization of ADR in what it terms Environmental Collaboration and Conflict Resolution (ECCR). In the 2017 fiscal year, the EPA reportedly used ECCR in 142 cases. That same year, the CPRC was recognized as among the "Top 25 Programs for Innovations in American Government" by Harvard's Ash Center for Democratic Governance and Innovation.

The U.S. Department of Justice has reported that it has identified approximately 250 cases in which an ADR process, typically mediation, either has been used or has been considered as a mechanism for facilitating settlement. Some examples include:

- **U.S. v. Martin:** A CERCLA cost recovery case where a judge, appointed as a mediator, conducted a one-day mediation that required each party to present its case in a concise fashion. The judge and the parties then met in private sessions until a successful allocation agreement was reached.
- **In re Mountain Water Rights Adjudication:** A local rancher mediated a case involving a grazing association and the U.S. Fish and Wildlife Service. The rancher held a series of meetings with the parties, other stakeholders, state and local agencies, and sporting groups to reach a successful agreement.
- **U.S. v. Farwest Fisheries:** A mediator was used to work with the parties to attempt to reach a settlement of an asbestos penalty case.
- **National Oilseed Processors Ass'n v. EPA:** A Court of Appeals mediator successfully identified a way for the parties to resolve their dispute in the appellate court.
- **U.S. v. 2.1 Acres:** A Court of Appeals mediator also assisted the parties of this case to reach a compromise (Schiffer and Juni 1996).

Experience in these matters, both before and after litigation has commenced, has shown that mediated negotiations encourage settlement as they tend to focus more on resolving the real issues separating the parties. Negotiations take less time and energy and are less expensive than litigation. The accuracy of the result of litigation is often dependent on the quality and the zeal of adversarial presentations (Dauer 1994). Mediation is a preferable dispute resolution process because the parties can reach agreement more quickly by identifying common interests and tailoring them into settlement options to meet their specific needs. Instead of the adversarial process of litigation, mediation allows the parties to reach their settlement objectives efficiently, to minimize the damage to their ongoing relationship, and to alleviate cost and delay.

In the environmental arena, mediation is particularly appropriate to avoid costs, delays, and the uncertainty regarding ultimate results. The following are considered by the U.S. Department of Justice as reasons why ADR may be helpful in resolving environmental disputes:

- **The ability of a mediator to conduct frank, private discussions may improve the outcome**. For example, some litigants may be loath to "put the cards on the table" in front of other litigants. A mediator may be able to convey information to other parties in an indirect fashion.

- **The range of issues is broad enough, or can be creatively made broad enough, to allow trade-offs and creative generation of options, especially when some options cannot be ordered by a court**. For example, in a National Environmental Policy Act dispute, underlying resource management decisions are likely the crux of concern, but cannot be reached by a court. By addressing concerns regarding the underlying dispute, a mediator may be able to fashion a resolution for the issue at hand. This result may forestall future litigation.

- **A mediator may be helpful in facilitating negotiation by breaking through impasses**. Such impasses may develop because of conflicts within interest groups, technical complexity, or uncertainty or political visibility or poor communication among the participants due to personalities or past history. For example, a mediator can defuse tension with a citizens' group concerned about a particular agency project by presenting negotiating proposals from all sides in an even-handed manner. In a case in which an impasse involves technical complexity, a mediator or a joint expert might even offer technical expertise on a given issue.

- **A thorough exchange of information may improve the outcome**. For example, a mediator can help to ensure that all issues are addressed, and that the heat of negotiating has not caused the parties to overlook an item that may be crucial to settlement implementation.

- **The participation of parties not directly involved in a legal action is necessary or beneficial to the settlement**. For example, numerous citizens' groups may be interested in a particular agency project; addressing the concerns of only the group that sued may be shortsighted, and invite future litigation from others.

- **The parties and issues are numerous, such that a facilitated, structured settlement process would be helpful, and no party is willing**

or able to take the lead role in establishing such a process. For example, CERCLA allocation disputes often involve multiple parties and issues. A mediator who provides a structure for allocation can assist the parties in reaching a global settlement (Schiffer and Juni 1996).

Certainly, businesses that are interested in resolving environmental disputes in a more efficient and cost-effective manner should consider mediation as a device to eliminate needless expenses and to maximize opportunities for a permanent solution.

9.2 Reasons for Foregoing the Use of ADR

Why then do so many businesses and their counsel frequently forego using mediation to resolve environmental conflicts? There are many reasons. ADR is a relatively new concept in the environmental field. *Black's Law Dictionary* did not even contain a definition of ADR until its 1990 edition. In the last 30 years, federal and state courts throughout the country have adopted mandatory ADR—mostly mediation programs giving litigants the opportunity to resolve lawsuits while avoiding the procedural and discovery delays that occur during the trial of the dispute.

In civil cases, ADR in general and mediation in particular will not work if senior corporate managers have a philosophy that winning is the only thing that matters (Carver and Vondra 1994). The common phrase for this type of approach in massive civil cases is "bet-the-company" litigation. Companies like Chevron, Toyota, and AT&T Global Information Systems were able to successfully integrate ADR techniques into their business operations to avoid litigation generally as their managers adopted the ADR principles wholeheartedly. As long as companies view ADR as an alternative, rather than the primary or preferred method of settling disputes, companies will not obtain sufficient experience with it to make significant and necessary changes in how they litigate cases (Carver and Vondra 1994). ADR is sometimes handled in the same way as litigation when the opponents "waste prodigious quantities of time, money and energy by reverting almost automatically to the habits of litigation" (Carver and Vondra 1994). Despite shortcomings and concerns, ADR is now being embraced by courts, litigants, and companies, large and small, throughout the U.S. when people want cost-effective, efficient, and relatively quick outcomes.

Some litigants simply trust judges to make the right decision rather than rely on mediators to bring parties to a negotiated agreement. Environmental cases are most frequently decided by judges after bench trials and rarely by jurors. CEOs and senior managers trust judges will be fair to their cause and not overly penalize their businesses for unintentional environmental mistakes. Litigation also buys time. It allows senior managers to take the opportunity to make better reasoned decisions involving the business's resources. One environmental lawyer has noted that the passage of time can result in a potentially fairer liability scheme, more lenient cleanup standards, more reasonable agency personnel, and perhaps even improved site conditions through natural attenuation of the contamination (Dean 1996). Sometimes, if the environmental conditions are stable and there are no adverse health consequences involved, there may be no reason to make a decision until all of the facts and circumstances are known. A premature decision to resolve a dispute without full knowledge of the consequences can be severely prejudicial to all the parties involved. In some cases, corporate managers need to maintain control over the environmental problem, the nature and timing of the cleanup, and the disclosure of contamination to the government so they can make correct decisions involving the financing and timing of remediation of environmental contamination. For these reasons they may not wish to agree to an expedited settlement using a mediator before there is an adequate factual basis to make a reasoned decision, or before a claim has fully matured.

Mediation is best suited for situations where there is a reasonable likelihood of compromise. This does not mean that the parties to a dispute think there is a reasonable likelihood of compromise but rather that the circumstances may allow such compromise. Parties in the throes of a dispute may not be able to accurately assess each other's interests in early settlement without assistance from a neutral party. The chances of success increase significantly as the parties learn to trust each other and identify mutual interests that can be achieved through a negotiated settlement. Mediators develop special skills to encourage this to happen. The EPA has had much success using mediation in environmental matters involving a large number of businesses with potential liabilities, which would mean they must eventually reach a consensus on how to allocate their share of cleanup costs to settle their dispute with the government.

Another factor influencing the corporate decision as to whether and when to participate in a mediation is the uncertainty that occurs in most environmental cases. Both the state environmental enforcement agencies and the EPA have an extraordinary amount of regulatory authority concentrated in a few individuals who have the ability to charge companies and individuals with environmental violations. This starts a process that is difficult to control and can inflict enormous burdens and financial damages on both individuals and businesses. The EPA, the state environmental protection agencies, environmental groups, and business leaders have been wrestling with conflicting theories of what works best to address environmental contamination, to minimize future impacts of business, and to deter bad actors from noncompliance. Is it more effective to increase enforcement of the federal, state, or local environmental laws, or to provide businesses with greater incentives to employ environmental management systems to prevent environmental violations and ensure swift cleanup when noncompliance occurs? In the face of this ongoing debate, many businesses are unable to tell when they should proactively use alternatives to litigation to resolve disputes. When a business's continuing existence is threatened, it may have no other choice but to use all available resources to defend itself where liability is so sweeping and the future is entirely uncertain. This dilemma can also result in "bet-the-company" litigation.

9.3 Deciding Whether to Use ADR or Litigation

In determining whether to proceed with litigation or mediation after a dispute arises, it is important to weigh the merits of both objectively. Litigation and mediation are not necessarily mutually exclusive. There are circumstances in which multiple strategies can be developed using mediation at a later stage of the proceedings. This may occur after discovery of pertinent facts or by using a summary jury trial procedure to assess the strengths and weaknesses of a case to determine whether a favorable settlement is possible.

Factors that generally favor litigation include (1) when there is a need for a binding, enforceable, and final judicial decision; (2) when procedural safeguards are necessary to improve the truth-seeking nature of the process; (3) when a fact-finding process is critical to resolution; (4) when an appeal system is necessary in the event of an adverse decision; (5) when established norms (i.e., principles of law and case law) are important to the outcome; and (6) when the outcome will further corporate goals, resolve uncertainties, or establish the rights and obligations of parties as a precondition for further legal proceedings (Dauer 1994).

In determining whether to proceed with ADR, it is important to analyze future transactional costs for planning purposes. Accountants and management consultants can analyze these transactional expenses, including legal and consultant fees, to determine how much each stage of the proceeding should cost and whether other alternatives to litigation could be more cost effective for future scenarios. Each situation is different and requires independent assessment of various techniques that can be used to resolve problems expeditiously and cost effectively.

A case involving the Rocky Flats nuclear weapons factory near Denver, Colorado, offered an opportunity to use ADR to cut down legal fees and costs in resolving a complex environmental dispute. Former Rocky Flats operators Dow Chemical and, later, Rockwell International defended a toxic tort class-action suit brought by 40,000 neighbors of the facility who alleged that health threats decreased their property values. Legal titans on both sides, including one plaintiff's lawyer from Cincinnati known as the "Master-of-Disaster," squared off for a classic legal battle. Approximately 16 years after litigation started, the plaintiffs were awarded $926 million in economic and punitive damages and a 10-year-long cleanup began, costing them $7 billion. The cleanup was completed in 2006–7, and in September 2018 Rocky Flats National Wildlife Refuge opened to the public. Public concerns and protests persist to this day. The case raises important public policy issues about how to resolve complex environmental disputes, particularly where the government may be paying all defense costs because of indemnification agreements it had with the ex-operators.

How might mediation have helped resolve the Rocky Flats case? Both sides needed to realize that jury verdicts, in most cases but particularly toxic torts cases, are wildly unpredictable. What is absolutely certain is the incurrence of extraordinary financial costs associated with bringing such a massive case to trial. For all parties, expert fees can range in the millions of dollars, even if the case does not go to trial. In addition, the U.S. Supreme Court and lower courts have made it more difficult to introduce expert testimony, especially where an expert opinion is not peer reviewed or subject to prelitigation scrutiny in a publication, or is unable to be independently verified by another expert following the same "methodology."

On the other hand, the defense in Rocky Flats was also extraordinarily expensive; at least $20 million was spent handling preliminary discovery alone. All parties would have benefited from using a mediator or a team of mediators to reach a final consensus without years of litigation.

Legal-risk reduction requires planning and cooperation from all involved. A consensus-based approach to plan how to handle disputes when they arise serves to reduce financial and human resource impacts on the business. It is important for managers to understand that there are innovative techniques to be used to resolve disputes without incurring unnecessary costs. It is equally important for government officials to understand that it takes a cooperative effort between business and government to reduce environmental risks.

9.4 Avoiding the Risk of Committing an Environmental Crime in the 21st Century

In the first decade of federal criminal environmental enforcement (1982–92), 232 companies of all sizes were convicted. Leading the list of companies that were charged were Exxon, Texaco, Nabisco, Ralston-Purina, Keebler, W.R. Grace & Co., Ashland Oil, Orkin Exterminating Co., Ocean Spray Cranberries, and Pennwalt (Thornburgh 1991). The prosecutions in these cases sent a message that was heard by the corporate community: environmental crime was bad for business, consumers, the general public, and the environment.

In the 1990s, companies were being prosecuted at a ratio of 4 to 1 over individuals. By 1997 this ratio reversed. In later years, Fortune 1000 companies were nowhere near as vulnerable to prosecution for environmental crimes as they were in the late 1980s. There are several reasons for this major change. Large companies were the first to institute early warning systems to prevent and detect violations of environmental law. Legions of articles were written about avoiding environmental crime, catching the attention of corporate managers with titles such as "Doing Time for Environmental Crime"; "Behind Bars: Prosecutors Sting Corporate Executives"; and "Environmental Crimes and the Sentencing Guidelines: The Time Has Come ... But It Is Hard Time." In 1991 Attorney General Richard Thornburgh of the Bush Administration, at an address to an environmental law enforcement conference, described polluters as follows:

> We are dealing with offenders who do some of the dirtiest work ever done to human health and the quality of life. They illicitly trade in sludge, refuse, waste and other pollutants, and they pursue their noxious concealments only for the sake of gain. Everywhere—on our land, in our water, even in the air we breathe—they leave their touch of filth. (Thornburgh 1991)

Thornburgh's strong words had the support of the public. In one national poll in the 1980s, the public ranked environmental offenses just below murder, but above heroin smuggling, skyjacking, and armed robbery (Bureau of Justice Statistics 1984).

Large companies listened to their environmental managers and their environmental counsel, who heard the beat of the government's environmental war drums and took preventive action. These same companies took advantage of the government's enforcement policies that favor audits and disclosures of noncompliance in return for mitigation of civil fines and penalties, but no criminal charges. Early alarms regarding prosecutors running amok using broad environmental statutes to ensnare the innocent turned out to be just that—alarms that woke up executives who had no desire to turn in their pinstripes for prison stripes.

Environmental criminal law, despite years of widespread and well-publicized enforcement activity, remains a tricky area where an unintentional mistake or accident can lead to adverse consequences, including a possible megafine. A corporate environmental management system must include a mechanism to reduce the risk of violations and mitigate overall civil and criminal liabilities. To understand how to create an effective environmental management system that includes a risk reduction strategy, it is necessary to know exactly what constitutes an environmental crime. The following is a summary of the criminal provisions of the four federal environmental statutes used most frequently by the government to prosecute environmental crime, to explain what the government thinks is an environmental crime.

9.4.1 The Resource Conservation and Recovery Act (RCRA)

Violators of the provisions of RCRA are subject to criminal penalties if the acts are committed knowingly. Any person who knowingly treats, stores, or disposes of hazardous waste without a permit, or who knowingly transports or causes to be transported a hazardous waste to an unpermitted facility, is subject to a maximum term of imprisonment of five years and a maximum fine of $50,000 per day of the violation. A person who knowingly treats, stores, or disposes of hazardous waste in knowing violation of a permit, or who knowingly transports or causes to be transported a hazardous waste without a required manifest, is subject to a maximum term of imprisonment of two years and a maximum fine of $50,000 per day of the violation. A person who knowingly omits or gives false information in a compliance report can be jailed for two years and fined up to $50,000 per day of the violation. A person who knowingly destroys, alters, conceals, or fails to file compliance documents can be jailed for five years and fined up to $50,000 per day of the violation. The most severe penalty under RCRA is the knowing endangerment provision, which provides for a maximum prison term of 15 years and a fine of up to $250,000 for an individual or $1 million for an organization that violates any of the sections of RCRA and knows at the time that they thereby place another person in imminent danger of death or serious bodily injury.

9.4.2 The Clean Water Act (CWA)

The Federal Water Pollution Control Act, commonly known as the Clean Water Act, provides for the imprisonment of up to one year, or a fine of up to $25,000 per day of the violation, or both, for any person who negligently violates any of various sections of the statute dealing with discharges or disposal of oil and hazardous substances, or negligently violates permit requirements. If the violator acts knowingly, a sentence of up to three years may be imposed, and fines can reach $50,000 per day of the violation. Furthermore, the violator who acts knowingly, and at the time of the act is aware that he or she thereby places another person in imminent danger of death or serious bodily injury, faces a maximum of 15 years in prison and a maximum fine of $250,000. Businesses can be fined up to $1,000,000 under this knowing endangerment provision. Second convictions double the maximum punishment available for all violations.

9.4.3 The Clean Air Act (CAA)

Under the Clean Air Act and its 1990 Amendments, a person who knowingly violates any of several sections of the Act may be imprisoned for up to five years, or fined, or both. Knowingly making false material statements, failure to notify or report as required, or tampering with or failing to install monitoring devices as required is punishable by a maximum of two years in prison, and a fine. Knowing failure to pay a fee owed to the U.S. under the Act is punishable by up to one year in prison and a fine. Similar to RCRA and the CWA, the CAA also contains a "knowing endangerment" provision, which punishes violators for known releases into the ambient air of any listed hazardous air pollutant, which the violator knew at the time thereby placed another person in danger of death or serious bodily injury. Knowing endangerment is punishable by a maximum 15-year prison term and a fine. If such a release is perpetrated negligently, however, the maximum sentence is one-year imprisonment and a fine.

9.4.4 CERCLA

The Comprehensive Environmental Response Compensation and Liability Act (CERCLA) imposes criminal penalties on persons who fail to report a release of hazardous substances, store hazardous substances, without notifying the EPA, destroy records concerning hazardous substances, or file false information in a sworn claim against the Superfund. A person who is convicted of violating these provisions may be sentenced to three years' imprisonment, or five years in the case of a second conviction, and receive fines in accordance with Title 18 of the Federal Criminal Code. Under Title 18, the government can prosecute persons who make false statements to the EPA and government investigators in violation of 18 U.S.C. § 371 (conspiracy) and 18 U.S.C. § 1001 (false statements). The penalties for these offenses are five years' imprisonment and up to $10,000 in fines.

Unlike bank robbery, skyjacking, and murder, environmental crime sometimes means entirely different things to different people. Compare these two descriptions of the same case that was prosecuted by the state of Ohio in the early 1990s. In *Ohio v. Stirnkorb*, the state convicted the operations manager of a hazardous waste landfill for violating Ohio's hazardous waste laws by unlawfully and recklessly failing to evaluate excess rainwater on top of a waste cell before ordering it pumped into an adjacent creek, an unauthorized location. The following descriptions were by a former environmental prosecutor, now a defender, and by the environmental prosecutor who handled the case, respectively.

The former environmental prosecutor said: the manager was acting during an emergency caused by a downpour, took measures to pump only clear water into the creek and colored water into holding ponds, acted under obscure and conflicting regulations, and no pollution of the creek was ever shown (Gaynor et al. 1993).

The environmental prosecutor said: the case of *Ohio v. Stirnkorb* shows how seriously the state of Ohio takes its environmental protection responsibility. This was not an accidental violation caused by an imprudent reaction to an emergency. The hazardous waste facility managed by the defendant had a stormwater management plan approved by the Ohio EPA. The plan required that stormwater be pumped to a holding pond, where it could be tested to determine whether it was contaminated. The defendant admitted he knew of the plan and had always used it previously. Stirnkorb also admitted he made a conscious decision not to use the method at the time for which he was cited. The only reason he gave for this choice to disregard the plan was that it was quicker to just pump the waste into a drainage ditch that emptied into a nearby stream (Muchnicki 1993).

The *Stirnkorb* case and several other early cases caused debate as to whether prosecutors had too much discretion in charging environmental offenses. In the past several years, however, environmental cases are becoming more routine as both prosecutors and defenders have become more accustomed to handling these types of cases. With more cases being filed over the last two decades, there are more decisions handed down by courts interpreting statutes and imposing prison sentences and hefty fines. In the final analysis, what constitutes an environmental crime as we enter the third decade of the 21st century has become more predictable.

What are prosecutors looking for when they exercise discretion in charging an environmental crime? Keep in mind that throughout the history of environmental enforcement, criminal prosecutorial decisions have been made on a case-by-case basis and are most often controlled by specific factual, evidentiary, and legal considerations. Prosecutors generally will want to review the nature of the wrongful conduct, the impact on the general public, the motive of the potential defendant, the criminal and/or regulatory history of that individual or company, and the effectiveness and deterrent value of a criminal versus a civil option.

Four factors are considered in determining the effectiveness of the criminal enforcement option: deterrence, public protection, retribution, and remediation. In addition, one seasoned environmental prosecutor, Kent Robinson of Portland, Maine, has said that each case has to have a major impact on the environment. Criminal cases are not prompted by technical violations of those trying to implement complex regulations. Most cases involve deceit of some kind, such as lying to regulators. "There's no quicker way to get a criminal violation than to lie to regulators. It shows criminal intent."

9.5 Enforcement of Environmental Laws

In July 1996 the EPA released its annual enforcement report titled *Enforcement and Compliance Assurance Accomplishments Report FY 1995*. This report contains a summary of each of the criminal environmental cases brought by the federal government or resolved by a plea of guilty or a conviction following a trial for the most recent fiscal year. We compare this report (featured in the first edition of this book) with the scenario in 2016 to determine what the government considers to be wrongful conduct that should be prosecuted criminally and to show how the government has ramped up enforcement. In 1996, the report provided information on how often environmental crimes are prosecuted in the federal system. It did not give summaries of state prosecutions, which vary greatly in number from state to state. Some states like California have had a substantial number of environmental prosecutions and are more active than other states that have prosecuted fewer cases. In 1996, the EPA reported that 70% of all enforcement actions are undertaken by states, and state penalties for violations are "orders of magnitude" higher than those of the federal government. Let's look back at 1996.

The EPA levied $76.7 million in fines in 1996 according to an enforcement update released on February 25, 1997. Iroquois Pipeline Operating Company paid the lion's share of the fines, $22 million, for clean water violations in building a pipeline. According to the EPA, this figure was the second largest in history. The largest was the $ 1 billion fine levied against Exxon in the Exxon Valdez oil spill case. Steve Herman, EPA's Assistant Administrator for Enforcement and Compliance, stated that "strong enforcement underscores [the Clinton] Administration's commitment to protect our air, our land, our water, and our health."

Only about 74 criminal cases were cited in the 1996 enforcement report, or about an average of 1.5 cases per state per year. The EPA did not list the cases that were either dismissed by judges or resolved by not-guilty verdicts by juries. Five of the cases summarized are simply updates of enforcement actions that were reported in earlier editions of the yearly enforcement report. The remaining 69 cases are a relatively low number, considering the amount of waste that is produced and disposed of each year in the U.S. The *Accomplishments Report* demonstrates that the federal government's environmental prosecutions, despite its rhetoric, are generally few and far between.

Now let's look back at 2016:

On December 19, 2016, EPA's Office of Enforcement and Compliance Assurance released its *Fiscal Year 2016 EPA Enforcement and Compliance Annual Results*. This was a banner year for the Agency because two extraordinarily large cases, BP's Deepwater Horizon spill and Volkswagen's Clean Air Act violations, dominated the enormous fines and penalties paid to the government by these companies' actions. But, let's look at the statistics in any event and compare them to 1996:

- EPA's enforcement action required companies in 2016 to invest more than $13.7 billion in programs, infrastructure, and equipment to control pollution.
- The Volkswagen settlement, approved by a federal court in early 2017, added an estimated $14.73 billion in injunctive relief.
- In 2016, EPA received $5.8 billion in federal administrative and civil judicial penalties.

- In 2016, EPA forced companies to commit to treat, minimize, or properly dispose of 62 billion pounds of hazardous waste.
- Further reductions of releases of pollution amounted to 324 million pounds that year.
- Private-party Superfund cleanup commitments that year exceeded $1 billion.
- EPA obtained commitments for the remediation of an estimated 17 million cubic yards of soils and an estimated 174 million cubic yards of contaminated water.
- EPA conducted more than 13,500 inspections and evaluations, initiated more than 2,400 civil and administrative cases, and concluded nearly 2,400 civil judicial and administrative cases all in 2016.

EPA's top enforcement highlights for 2016 included the following cases:

At the beginning of FY 2016, Volkswagen agreed to a $14.7 billion settlement regarding allegations of cheating emissions tests and deceiving customers on vehicle fuel performance. Volkswagen agreed to buy back over 500,000 cars and provide additional compensation to consumers. This cost was estimated to be $10 billion. Another $2.7 billion was allocated toward environmental remediation and $2 billion toward promoting zero emissions vehicles.

British Petroleum's Deepwater Horizon oil rig exploded in the Gulf of Mexico on April 20, 2010. Four million barrels of oil flowed into the gulf over the course of 87 days until the flow was capped. The company agreed to pay a $5.5 billion CWA penalty as well as $8.8 billion for natural resource damages. This does not include the numerous private-party settlements for medical claims and economic damages.

Duke Energy entered a $3 million cleanup agreement with the EPA following a large coal ash release in the Dan River in February of 2014. The agreement also required continued monitoring of surface and drinking water as well as sediment assessments.

In January of 2017, the City of Tyler, Texas, agreed to upgrade its sanitary sewer system to resolve CWA violations and pay $563,000 in civil penalties. The City's alleged violations include "frequent discharges of raw sewage to waters of the United States and failure to prevent sanitary sewer overflows" through inadequate operations and maintenance.

In June of 2017, home developer NVR Inc. reached an agreement with the EPA to implement a program to improve compliance after EPA inspectors identified inadequate sediment and erosion runoff controls. The company also agreed to pay a $425,000 civil penalty.

In June of 2017, Rocky Mountain Bottle Company reached a settlement with the EPA after violating the CAA by failing to submit an application that included applicable requirements in the company's Title V permit. Further, the company did not obtain pre-construction permits or install pollution control equipment prior to conducting a furnace expansion project. The company agreed to implement an emissions reduction system and an emissions monitoring system. The company agreed to pay a $475,000 civil penalty.

In September of 2016, Occidental Chemical Corporation entered into a settlement agreement with the EPA over CERCLA violations. The company was found responsible for pollution discharges from a Newark pesticide manufacturing plant from the 1940s to the 1960s. Hazardous substances were found in the soil and groundwater of the old plant. All of this was also found in the lower Passaic River. The company will oversee a cleanup project of 8.3 miles of the river. This project is estimated to cost $165 million.

In January 2017, the EPA settled with Innophos to resolve RCRA violations. The company was discovered to be sending toxic waste streams to a neighboring facility that was not permitted to receive them. The company also failed to submit annual reports to the Louisiana Department of Environmental Quality. Innophos agreed to cease shipments of toxic waste to any facility that is not permitted to receive it as well as pay a $1.398 million civil penalty.

In January 2017, the EPA reached a settlement with Bandit Industries, Inc. after the company allegedly violated the CAA 2,552 times by selling diesel engines not covered by the certificates of conformity required by the act. The company faced no injunctive relief but was made to pay a $3 million civil penalty.

In September of 2016, Rutgers Organics Corporation and the EPA settled on an agreement that will cost the company $20 million. The company's chemical manufacturing plant in Ohio was responsible for soil and groundwater pollution as well as contamination that eventually reached Little Beaver Creek. The company will be responsible for cleaning the work site as well as sections of the creek. In addition to this, the company will set up a trust fund to protect local drinking water sources and also reimburse municipal agencies $1 million for their response and assessment costs.

Final statistics in 2016:

2016 EPA Environmental Enforcement Statistics Administrative penalties assessed: $44,000,000

Civil judicial penalties assessed: $5,746,000,000

Fines and restitution: $207,000,000

Injunctive relief: $13,700,000,000

Superfund cleanup commitments: $1,149,000,000

Criminal environmental cases usually involve egregious actions by individuals who have decision-making authority, and hence, control, over the disposal of regulated waste. The government highlights the most inculpatory facts and avoids mentioning the defenses in these case summaries, so it is difficult to tell how many of the defendants were under the mistaken belief that their actions were legal or were simply uninformed. The federal criminal environmental statutes require low levels of criminal intent necessary to support a criminal conviction. People who handle toxic wastes and pollutants are assumed by the courts to know the consequences of unpermitted discharges of such wastes into the nation's ecosystems. Prosecutions are invariably successful when people discard noxious wastes into the environment without a permit. The low standards of culpability have caused considerable angst among legal commentators and defense counsel, but they have been upheld repeatedly by the courts mainly because most environmental crimes involve outrageous facts.

For centuries a fundamental tenet of jurisprudence in the English-speaking world traditionally has been the requirement of establishing the knowledge of the defendant as a key element in a criminal proceeding. A noteworthy exception to this general rule has evolved judicially in cases involving the violation of statutes that were created in the public welfare. Although the public welfare doctrine originated in two cases dealing with the Federal Food and Drug Act, in the past decade courts have expanded the doctrine to include the various federal environmental statutes. In practice, this expansion has resulted in the reduction, or even elimination, of the requirement of actual proof of criminal intent in successful environmental prosecutions.

Several of the major environmental statutes (including RCRA, the CWA, and the CAA) explicitly include a knowledge requirement for a violation to occur. A question that has arisen frequently in litigation under these statutes, and in scholarly commentary, is whether the explicit statutory language used by Congress should supersede the common-law public welfare doctrine. Most courts that have addressed this issue have concluded that the congressional history of the statutes prevents the outright imposition of strict liability for most environmental violations. The same courts, however, permit a lessening of the direct knowledge or intent requirement and allow juries to draw inferences of knowledge from the facts established. Courts have greatly assisted the prosecution of environmental crimes by broadly construing the public welfare doctrine so that businesses with any impact on the environment are within its reach.

The second major weapon available to the government in environmental prosecutions is the responsible corporate officer doctrine, which is inextricably intertwined with public welfare legislation and is a formidable tool to attach criminal liability to people that law enforcement agents, and others, euphemistically refer to as "higher-ups." The responsible corporate officer doctrine can effectively place criminal liability upon officers, directors, and any managers at any level of the company who know, or by virtue of their position have reason to know, that a violation has occurred or is likely to have occurred. The responsible corporate officer need not have directly caused, ordered, or even been involved in the conduct giving rise to the violation in order for personal criminal liability to be attached. That person need not, in fact, have any direct knowledge of the violation at all.

The scholarly debates over the reach of the environmental statutes have not subsided as the government has increased its efforts to prosecute environmental crime. A close look at the actual cases the government has prosecuted shows, however, that despite the debate, the actual criminal conduct has been very clear and the defendants, remarkably guilty. Frequently, the disposal is so blatantly illegal and obviously deleterious to the ecosystem that defendants plead guilty simply to avoid an unnecessary trial and throw themselves, literally, at the mercy of the sentencing court. These cases generally involve deliberate disposal of hazardous wastes in rivers, oceans, wetlands, sewer systems, and fields, and on abandoned properties. On occasion, there is a significant health hazard to people, in addition to severe damage to the ecosystem, created by the discharge of the waste. Penalties usually, though not always, increase with greater environmental impacts. Frequently, the disposal is accompanied by false statements made knowingly to the EPA and state environmental protection agencies by businesses or their environmental consultants. Environmental crime is occasionally committed in conjunction with other crimes, such as racketeering, mail fraud, or simple fraud. Also, environmental crime occurs when businesses adopt poorly conceived disposal plans.

9.6 Practical Steps to Use Systems to Manage the Environment

Would an environmental management system have improved the performance of these businesses or employees and mitigated the possible violations? To improve environmental performance and minimize environmental risks and exposures, businesses need to understand the direct and indirect consequences of their actions. The government reports that one company was warned by state officials not to dump its toluene waste in its trash dumpster prior to the deaths of two small children who were asphyxiated by the chemical fumes. That company could have altered its disposal practices if it had known the consequences of its actions. Yet, the business apparently was unwilling to act merely on the basis of a warning from state officials. It took the chance that its disposal methods would not cause a physical or environmental injury. This is the deliberate risk-taking behavior that often results in negative and severe consequences for the company.

A systems approach to the minimization of risk could have provided the business with a range of possible consequences of the business's actions, including civil and criminal regulatory violations. Managers overseeing the disposal of waste could have prevented these accidental deaths by employing a properly implemented environmental management system. The system has to have the capability of internal controls to work effectively to eliminate noncompliance. Performance must also be independently reviewed to assure no breakdowns in the system. Risk reduction environmental management systems have to be dynamic, as well, to change with new circumstances and problems faced by the business each day.

It appears from the government's descriptions of cases in its enforcement report that management actively participated in the illegal disposal in many of the cases cited in the report. Again, the defenses and mitigating statements made by the defense to the jury during the trial or to the judge at sentencing are not included in the report. Nonetheless, an environmental management system is entirely useless if it will not be embraced by committed members of the management team. Businesses that have regulatory responsibilities need to have a carefully defined program that weeds out managers and employees who disregard the law, regardless of the consequences to business operations.

Law enforcement authorities rely on tips from neighbors in discovering environmental crimes. Public citizens frequently make complaints about what they perceive as environmental nuisances. Employees frequently report on managers and fellow employees who they think are violating the law. These employees may have been directed to participate in unlawful activities. The reports are generally made to local or state officials, OSHA, the EPA, or the FBI. Law enforcement authorities take these reports seriously and have been increasing the number of trained agents to investigate these types of tips. Since the Pollution Prevention Act of 1990, the numbers of federal environmental agents have been steadily increasing.

Employees generally notify authorities because they want "to do the right thing," to clear their own conscience if they were involved, or because they believe that their own inaction may be perceived as complicity with the illegal actions of the business. Employees also report on their businesses when they become disgruntled, often as a result of actions taken against them when they are sexually harassed, discriminated against on the basis of race, gender, or sexual orientation, fired, demoted, or passed over for promotion. These employees become another set of eyes and ears of the regulatory authorities. An environmental management system can prevent violations of environmental law in the first place. Another important function, however, is for the system to serve as a reporting mechanism to responsible managers when incidents occur. If properly reported, documented, and explained, an employee may not perceive the event as an environmental crime. This is an effective way to avoid false or mistaken claims by employees against a business that is committed to detecting and correcting noncompliance.

Many of the EPA's criminal cases involve sole proprietorships, where the illegal action is taken by the business's sole employee. An environmental management system can be designed for the smallest businesses, including those that have only one employee. The program would amount to a checklist for an individual to follow in conducting daily business operations. From a small gasoline station operator or a family-owned dry cleaner to the largest multinational corporation, all businesses need some guidance in managing environmental risk and minimizing environmental impacts. Perhaps none of the individuals mentioned in the EPA's enforcement report would have altered their conduct on the basis of an environmental management system. However, given the alternative of spending months, if not years, in jail for senseless or stupid acts, even the most hardened and malignant midnight dumper perhaps would have some misgivings about perpetrating an obvious crime on the rest of society.

What about those businesses whose employees lie in reports to management and to regulatory authorities? Businesses and their environmental consulting firms have been prosecuted, as have environmental laboratories, for falsifying data and other information. Extensive computer networks, management oversight, and regulatory review are necessary to discover the false statements in official and unofficial documentation. It is also necessary to find ways to prevent these people from violating the law.

Incipient behavior often causes problems, like impossible sales goals that result in salespeople bribing purchasing agents to achieve those goals. To avoid creating unnecessary risks, like guaranteed and unrealistic delivery times that inadvertently cause accidents, management systems need to create positive incentives that foster good corporate citizenship and remove barriers to compliance that cause problems and accidents to happen because businesses will be held responsible for failure to exercise supervision over work performed on their behalf.

9.6.1 Implementing the Systems Approach

The systems approach to environmental management is a method to manage and reduce environmental risk. We introduced this concept in 1997 with the first publication of this book. At that time, as now, it was obvious that implementation of such a system could not begin before there was commitment from top management with accountability for continuing and active participation in promoting and overseeing the environmental management system. The systems approach, as we labeled it then, could be used by managers either to create or to enhance an existing environmental management system.

In 1997, we identified the most important concepts regarding how top management, including the Board of Directors, the CEO, and the C-level Suite could get involved with the environmental professionals at the company in the development and continuing operation of an environmental management system. We reviewed a total of 38 elements of a formal environmental management system. We also included principles of governance that could be used to provide a framework and a support system for managing environmental risks on a daily basis. In this chapter, we will revisit each of these 38 elements with specific examples of what companies were doing back then by employing voluntary strategies to reduce risks while at the same time effectively managing the system. The governance principles were added to show how business organizations could create structure and control of their systems. These principles came from the National Center for Preventive Law's (NCPL) corporate compliance project that was published in 1996. The NCPL Commission on Corporate Compliance identified 12 essential principles to adopt to create effective corporate management. This project included guidelines for creating an effective corporate compliance program. We applied these principles and guidelines to support the development of a robust environmental management system for those companies that impact the environment.

Since 1997, environmental management systems and good corporate governance policies and procedures have merged to form the modern-day corporate governance programs. Many companies now report their compliance achievements in Environment, Social and Governance ("ESG") reports. These reports are also commonly identified as "Sustainability" reports.

As we look back on the 1990s, when environmental management and compliance systems were just arriving, all of the standards, principles, considerations, and examples used back then are incorporated and, quite frankly, thriving in the ESG-Sustainability reports of 2018–19. This transition is not meant to convey that times have changed. Of course, they have. But, what is really meant by this comparison over the course of approximately 20 years is the original elements of a systems approach to environmental management in a governance context worked well then, and work even better today. Granted, many elements have been enhanced by more comprehensive ESG-Sustainability reporting. Companies have devoted substantial resources to combating climate change and achieving worthy social goals like diversity and inclusion. Their measurement of progress is contained in the ESG-Sustainability reports. A comparison of then and now shows a remarkable consistency in the values and effects of responsible commitment to protecting the environment.

9.6.2 Commitment from the Top

In 1997, we recognized that commitment from senior management to improve the environmental performance of the company was necessary whether a system is being designed pursuant to ISO 14001 or by other standards. Without this firm support, it was unlikely that efforts to implement an effective system would succeed. For this reason, we called for top management to get involved and to recognize that environmental performance is among the highest corporate priorities and act accordingly. We mentioned that the system must be created to reflect the business's culture, ethos, and corporate objectives (NCPL 1996). At this time, a company we highlight in this book, IBM, became the first major multinational to earn a single global registration of its EMS to ISO 14001 through its early management endorsement and commitment to the environment (IBM 2018, 7).

In 1996, the NCPL Commission found that the highest governing authority within the business should endorse the governance program and give it visible support. Many environmental policies back then fit within the governance system and began with a personal statement from the CEO, the Board of Directors, or senior management regarding the company's commitment to the environment. For example, David Simon, Group Chief Executive of British Petroleum, introduced BP's policy in March 1993 with his personal commitment to BP's health, safety, and environment policy. Current Group Chief Executive Bob Dudley continued this commitment by introducing the 2018 Sustainability Report. As a company, Johnson & Johnson introduced its new environmental leadership policy with a statement by Chairman and CEO Ralph Larsen to the shareholders. Alex Gorsky, Chairman and CEO of Johnson & Johnson, delivered the company's 2018 Health for Humanity Report to stakeholders with a high-level summary of the report. American Express began its *Environmental Principles* with a statement from the Public Responsibility Committee of the company's Board of Directors regarding how the committee will oversee management's commitment to environmental policies and practices.

Significant input, commitment, and leadership from top management continue to be necessary to make an environmental management system successful. Virtually every ESG-Sustainability report featuring environmental management starts with a pronouncement of support from the CEO. Management, we said in 1997, can demonstrate their interest and commitment to the environmental management system by reviewing and approving specific provisions of the program, making clear assignments of responsibility for the program, and holding themselves accountable for those compliance activities that they initiate or oversee.

Top management needs to do more than just set an example for the rest of the company to follow in protecting the environment and minimizing environmental impacts. The environmental management system should be designed with input from knowledgeable people throughout the business and from outside consultants if necessary. Open lines of communication with internal and external stakeholders can be established and maintained during this process so that the system is effectively created and received. Thoughtful environmental planning combined with product and process life-cycle analysis can increase the likelihood of achieving environmental performance targets as employees learn to manage and reduce their risks and liabilities.

Virtually all companies should have the resources to create an effective environmental management system. The program needs to be tailored and fine-tuned with specific regard to the size, form, complexity, and history of the business. Some companies' systems will be more formal than others. All should be in writing and set forth program definitions and operating practices in clear terms. The environmental management system should be readily available to all employees. Employees should easily understand environmental management system procedures and know where to get help or their questions answered. To assure fundamental fairness in the operation of the system, all employees subject to the system should be treated equally and consistently. Mechanisms should be written into the system to prevent retaliation for raising compliance issues.

It is important that the environmental management system leaves the choice of the format of the system up to the company itself. The company will set the level of compliance activity based on the company's existing policy, the totality of the company's environmental impacts, the size of the company, the type of business it conducts, its available assets, and its existing reputation and external relations.

9.6.3 How to Get Started

Most of the environmental management systems we have developed proceeded with an initial environmental review. The purpose of the review is to identify legislative and regulatory requirements; to identify the environmental aspects of the company's activities, products, or services; and to ascertain which aspects create significant environmental impacts, risks, and liabilities. This preliminary review should also be used to evaluate existing environmental performance. Relevant internal criteria, external standards, codes of practice, and sets of principles and guidelines can be collected to benchmark existing performance. The review should also consider existing policies and procedures that deal with other related activities, such as procurement, contracts, and third-party vendors and software systems. If there are any previous incidents of nonconformance, the review can include feedback from prior investigations, regulatory proceedings, and conflict resolution in the courts. The initial review can identify what functions or activities of the company's other systems enable or impede the company's environmental performance. Collecting and analyzing this information during the initial review can also serve the dual purpose of discovering opportunities for competitive advantage and as a baseline of existing views of stakeholders.

Information can be collected by submitting questionnaires to management, employees, and stakeholders; conducting interviews with key personnel; inspecting and measuring existing systems; reviewing pertinent records; and benchmarking. Government agencies can be contacted to identify compliance permits. Local or regional libraries or databases can be queried, along with industry associations, larger customer organizations, manufacturers of equipment in use, and other related businesses. Upon completion of the initial review, the company is ready to begin drafting an environmental policy.

9.7 Environmental Policies That Work

A corporate environmental policy will establish an overall sense of direction and set the principles of action for the business. The goals of environmental responsibility and performance required of the business will be judged against all subsequent actions. The following factors should be included or considered in the development of a sound environmental policy.

9.7.1 The Mission, Vision, Core Values, and Beliefs of the Business

Many companies, like Hewlett-Packard, stated in their early environmental policies that they are committed to conducting their businesses in an ethical and socially responsible manner. HP went further than most by acknowledging that it has an "aggressive approach to environmental management," which also included occupational health, industrial hygiene, safety management, and ecological protection. HP believed its approach is "consistent with the spirit and intent of our established corporate objectives and cultural values." Consistent with these core beliefs in its *Sustainable Impact Report* of 2018, HP's "vision is to create technology that makes life better for everyone everywhere." The CEO, Dion Weisler, states,

> "A growing number of our customers, consumers and employees are passionate about the environment and social justice, and the expert companies like ours to lead with purpose"

(HP 2018, 4).

ComEd begins its policy with the statement that it has long been committed to protecting the environment because of its special position as the major supplier of electrical energy in northern Illinois. It adds, like many other companies, a statement that it believes "our customers want and expect us to respect the environment." Further, ComEd makes clear that the company is a member of each community in which it operates. Thus, its employees are deeply interested in the effects its operations have on the environment and the health and safety in the service area. Like HP, ComEd states that it is taking "aggressive action" to protect the environment, and adds that it takes pride in its compliance record and cooperation with authorities. Exelon's 2017 Environmental Policy begins by acknowledging that "our ability to serve customers, create value for investors and contribute to our employees' and communities' shared expectations for the future directly depends on how well we manage our interactions with the environment." Their policy has an emphasis on risk management, and in fact commits to conforming with ISO 14001.

Graham Partners, in its 2018 Sustainability Report, commits that, "As investors in businesses, we care not only about financial returns, but also that our investments are having a positive impact on the world." The tone of the HP, ComEd, and Graham Partners policies immediately indicates that their policies reflect, incorporate, and integrate with their business's culture, ethos, and corporate objectives.

9.7.2 Environmental Responsibility and Leadership

Two closely related topics are environmental responsibility and leadership. To indicate responsibility, Bristol-Myers Squibb says it "will ensure that each employee understands the importance of, and is responsible and accountable for, integrating environmental health and safety considerations into their daily responsibilities." In his 2013 book *Conscious Capitalism*, John Mackey, the founder of Whole Foods, wrote that "true investors have an important purpose to their work and collectively create great value for society." Johnson & Johnson stated in the past that its goal is environmental leadership. In its 2018 Health for Humanity report, Johnson & Johnson stated that its approach includes "continuing to raise the bar for ourselves by setting five-year public commitments to both improve our performance and demonstrate leadership across environmental, social and economic topics relevant to our business." The report goes on to mention their long history of environmental leadership and mentions several programs designed to further develop environmental leadership in its employees.

In Graham Partners 2017 Sustainability Update, Sarah Kolansky, Associate Director of Sustainability, explains that "those of us in the Environmental, Social, and Governance investment world are reminded nearly every day about the long-term benefits of addressing these issues. Studies show that companies that actively embrace Environmental, Social, and Governance opportunities outperform those that do not."

9.7.3 Sustainable Development

In 1997, there was a growing trend for companies to endorse the concept of "sustainable development" in their policy statements. In that era, sustainability simply meant leaving future generations with a better world economically, socially, and environmentally. Xerox was one of the first companies to state it has an aggressive environmental management program that created products for a sustainable future. In 2018, with the publication of its Corporate Social Responsibility Report, Xerox's Chief Executive Officer, John Visentin, stated:

> We believe what's good for society is good for Xerox … . We recognize the World's challenges, like climate change, and human right(s) issues, continue to devastate our planet and we understand the role we play. Combined with corporate environmental initiatives, Xerox calls for the design and use of products that are manufactured, used, and reused in a manner to improve the environment.

In the late 1990s when Xerox first developed its environmental policies, significant transition was already underway from the old mode of business to electronic or networked offices and multifunction products with multiple life cycles. Xerox believed this change was vital for future sustainability.

Starting in the 1990s, Baxter stated it agreed to "strive to conserve natural resources and minimize or eliminate adverse environmental effects and risks associated with our products, services and operations." By 2018, Baxter's Corporate Responsibility Report, titled "Making a Meaningful Difference," emphasized sustainability in virtually every aspect of their business. Most especially, "We use Environmental, Health, Safety and Sustainability reviews and life cycle assessments to understand product environmental performance and requirements and drive ongoing improvements across the value chain" (Baxter 2018, 11).

9.7.4 Communications with Interested Parties

Virtually every company that we mentioned in the first edition recognized the need for open communications with interested parties. Environmental communications expert A.J. Grant back then pointed out that effective communications with external stakeholders cannot and should not begin before a company understands the message and delivers it effectively and meaningfully to internal stakeholders. The message must begin with an environmental policy that is soundly developed and available to all employees. Employees need to develop trust in management that the policy is meant to be implemented in a consistent, fair, and effective manner. At that point the policy can be made public.

In our first edition we stated that General Motors (GM) confirmed that it would continue to participate actively in educating the public regarding environmental conservation. In an effort to improve dialogue and community relations, in the 1990s Chevron's Oak Point Plant organized a community advisory panel composed of civic leaders and members of the community to discuss environmental issues and concerns. GM, like other companies back then, took affirmative steps to work with the communities in which their plants were located to encourage open communications regarding operation and safety risks.

GM's 2018 Sustainability Report, "Transformation in Progress," continues this tradition with "Collaborating to Move Humanity Forward" (GM 2018, 20). This 196-page report summarizes GM's vision for the future as "Zero crashes, zero emissions, and zero congestion," a company committed to safety for its customers, its communities, and the public (GM 2018, 4). This includes an "all-electric future." Nissan's "Sustainability Report 2018" contains the same message. President and CEO Hiroto Saikawa states,

> Nissan aspires to lead the world toward the realization of a zero-emission, zero-fatality society, at a moment of great transformation in the auto industry. We also see a significant opportunity to address unprecedented challenges facing the planet and humanity, including climate change, resource depletion, a high number of auto accidents and a widening socioeconomic divide. (Nissan 2018, 2)

9.7.5 Continual Improvement

Increasingly, policies contain a commitment to continual improvement, a term that picks up the concept from the Japanese term *kaizan*, which means small, ongoing, and continuous improvements (Mcinerney and White 1995).

Xerox adopted an Environment, Health, Safety and Sustainability governance policy in 1991. The roots of this policy were in "continuous improvement." Now that policy states: "We are committed to designing, manufacturing, distributing, and marketing products and processes to optimize resource utilization and minimize environmental impact" (Xerox 2019, 16). Westinghouse similarly declares it

> will design, source, produce, market and deliver our products and services in a safe, environmentally sound and socially responsible manner … . [This includes] [c]ontinually improving EHS management systems and performance by establishing meaningful objectives and targets, and evaluating EHS performance.

(Westinghouse 2017, 1)

It further states that it is dedicated to the concept of continuous improvement of its performance in environment, health, and safety. The Public Service Company of Colorado similarly states that it is continually searching for ways to improve its environmental and safety performance. Duke Power requires each employee to pledge to "continually look for ways to improve performance and better protect the environment." StorageTek states that it is striving for continuous improvement in pollution prevention, waste minimization, and resource conservation.

9.7.6 Pollution Prevention

In its 2019 Sustainability Report, 3M's CEO, Mike Roman, declared the company has prevented 2 million tons of pollution since it adopted the Pollution Prevention Pays program (3M 2019, 5). Since 2002, 3M has reduced its greenhouse gas emissions by nearly 64% and nearly doubled its revenue. Quite a story.

In its 2018 "Global Citizenship Report," Bristol-Myers Squibb stated, since 2009, "our sustainability goals have guided programs to reduce our environmental footprint. The goals encourage continued, year-over-year progress that has been a reduction in greenhouse gas emissions by more than 24% and water use by about 14%" (Bristol-Myers Squibb 2018, 20). "[B]y purchasing 29 million kilowatt hours" of renewable energy, BMS removed the equivalent of 4,621 passenger vehicles off the road for one year (Bristol-Myers Squibb 2018, 20).

Prevention of pollution was a key topic contained in all of the policies we reviewed in 1997, and a required element of policy conforming to ISO

14001. For example, Monsanto has pledged to reduce all toxic and hazardous releases and emissions, working toward an ultimate goal of zero effect. DuPont declared that it will "drive towards the generation of zero waste at the source." DuPont further committed to reuse and recycle materials to minimize the need for treatment or disposal and to conserve resources. Where waste is generated, it will be handled and disposed of safely and responsibly. By 2018, DowDuPont, now a merged giant, furthered both goals in the "Sustainability Report" with the tag line "Seek Together" and in DuPont's 2018 Global Reporting Initiative. Dow's 2025 sustainability goals include leading the blueprint, delivering breakthrough innovations, advancing a circular economy, valuing nature, safe materials for a sustainable planet, engaging for impact communities, employees and customers, and world leading operation performance (Dow 2018, 20). For DuPont, the goals are footprint reduction, embedding sustainability in the innovation process ("sustainable innovation"), transparency, and engagement (DuPont 2018, 21).

Johnson & Johnson has an impressive record of pollution prevention. It decreased reportable chemical releases from its U.S. manufacturing facilities by 50%. The company has eliminated ozone-depleting chlorofluorocarbons (CFCs) from more than 120 products and processes worldwide and upgraded thousands of its storage tanks to state-of-the-art standards.

9.7.7 Coordination with Other Organizational Policies (e.g., Quality, Occupational Health, and Safety)

Thoughtful companies integrate their environmental management system with business operations and other company compliance policies. For example, in the 1990s, Lockheed Martin agreed to integrate its environmental, safety, and health considerations into strategic business discussions, engineering design, procurement, facilities management, and production. In its 2018 Sustainability Report titled "The Science of Citizenship," Lockheed's Chairman, President, and CEO Marilyn Hanson states, "we know that sustainability reaches far beyond environmental protection efforts." The company seeks to provide advanced technologies that play a role in ensuring a sustainable future for all ... that is integrated with business integrity product impact, employee well-being, resource efficiency, and information security (Lockheed Martin 2018, 2). The Graham Partners 2018 sustainability program currently focuses on the areas with the greatest impact: energy usage and employee health and safety. By engaging directly with their portfolio companies, they are able to execute their purpose, while minimizing negative impacts on their people and our planet.

The environmental management system should be consistent with all other existing company policies. Each of the company's multiple compliance policies will typically identify important business objectives and goals. To ensure consistency, the environmental program can specifically incorporate portions of existing policies. The NCPL Commission describes a company, for example, that incorporated its existing code of ethics, vision statement, and corporate guidelines, as well as existing compliance activities, as part of its overall compliance program. An environmental management system is an integral part of a company's overall efforts to achieve business goals together with compliance objectives and reduce risks and liabilities at the same time.

9.7.8 Specific Local or Regional Conditions

Many companies' operations have specific impacts on local or regional conditions. Mining, paper production, and large-scale electronics come to mind. Some of these impacts can be positive. In our first edition, we noted BASF co-located plants so that the industrial waste created in one facility can be used as a fuel in another. This action stimulated the economy, improved the environment, and avoided the high cost of transporting and disposing of industrial wastes. Other companies have designed transportation programs including ride-share, van pools, and other transportation alternatives and have made other accommodations to employees including telecommuting in order to reduce environmental impacts. Some of these companies provide incentives for employees to find alternative means of transportation to and from the workplace. Nissan introduced a policy in which employees can freely choose to spend up to 40 hours a month working from home. In fiscal 2017, 6,300 employees took advantage of this system (Nissan 2018, 143). Local or regional planning initiatives can be incorporated into the environmental policy of the company to reduce impacts on ecosystems, and in the case of Nissan, promote an inclusive and diverse work environment.

9.7.9 Compliance with Relevant Environmental Regulations, Laws, and Other Criteria to which the Business Subscribes

Virtually all of the companies commit to compliance with relevant laws. This element is required by all formal EMSs. As an example, BFI states that it will comply with all applicable environmental health and safety laws and regulations. These actions are intended to minimize adverse environmental health or safety effects from the company's business activities to achieve a cleaner global environment. .

9.7.10 Minimize any Significant Adverse Environmental Impacts of New Developments through the Use of the Integrated Environmental Management Procedures and Planning

A successful management system can be tailored to the needs of a growing business that is experiencing increased regulatory responsibilities. While business is expanding, companies need to avoid developing environmental management systems out of proportion to a company's risks of noncompliance. Companies must focus on functions to achieve a proper balance. In the worker safety area, for example, the NCPL Commission found a risk assessment of operating activities might be completed to identify high-risk operations. These operations could then be addressed with greater attention and detail. Lower-risk operations, on the other hand, might be addressed with less comprehensive direction and monitoring. ISO 14001 requires a periodic review to realign resources and risks.

9.7.11 Development of Environmental Performance Evaluation Procedures and Associated Indicators

A company needs to develop methods to ensure that the environmental management system is working. Both the EPA and the states have been seeking to discover appropriate environmental indicators to let them know that their regulatory programs are achieving maximum benefits for reasonable costs. A company may consider the benefits of participation in voluntary programs that seek to embody a systematic approach to environmental protection and test alternatives to regulatory requirements. Similarly, a company needs to have indicators, such as a lack of environmental compliance violations or public interest group complaints, a continuous record of reduced waste and more recycling, or other measures to let it know that the system is working and objectives are being achieved.

9.7.12 Embody Life-Cycle Thinking and Product Redesign

ConocoPhillips states in its environmental policy that it encourages life-cycle assessments in the development of its products. HP has created its "Full Circle Approach" (HP 2018, 20). Lockheed Martin has a goal of generating $1 billion of life-cycle cost reduction from manufactured products related to the use of resources and impacts on human health and the environment (Lockheed Martin 2018, 25). This book is replete with examples of other companies that incorporate this thinking into daily business operations. These examples include redesign of products like cell phones, vacuum cleaners, cars, plastic bottles, electronics, pesticides, herbicides, tires, copiers, and printers to reduce their environmental impacts. Many companies have undertaken waste stream analyses that have changed production methods. By 2010, the 50th anniversary of its founding in a small farmhouse in rural Pennsylvania, Graham Engineering Corporation generated innovations in packaging, machinery, recycling, and building products that had enabled its legacy operating businesses to generate tremendous growth reaching over a billion dollars in annual revenue, largely by focusing on life-cycle thinking and product redesign. These studies have also resulted in product redesign to eliminate waste, as well as the use of toxic substances and unnecessary parts, so that products can have a second or third useful life and not be discarded into landfills until their usefulness is truly exhausted.

Environmental policies that endorse life-cycle analyses, coupled with sustainable development and continuous improvement, are on the cutting edge of environmental proactive thinking.

Xerox's 2018 Corporate Social Responsibility Report announced that "newly-launched, eligible" products satisfied both the Electronic Product Environmental Assessment Tool and the EPA Energy Star eco-labels. Xerox also developed and conducted full Lifecycle Assessments to evaluate environmental impacts at various points in the product's life cycle.

3M's 2019 Sustainability Report adopted the phrase "Science for Circular," that is, "[d]esign solutions that do more with less materials, advancing a global circular economy" (3M 2019, 5). 3M further stated that it "launches approximately 1,000 new products each year, crossing many industries and geographies, the impact will expand greatly with each successive year" (3M 2019, 5). Similarly, Xerox's 2018 Corporate Social Responsibility Report highlights its efforts to achieve a "circular economy" with consumables take back and recycling; using waste as a resource; and by designing products for the future. Waste Management's 2018 Sustainability Report, titled "Driving Change," has an entire section devoted to how this company has created a "Life Cycle Assessment Approach to Recycling" (WM 2018, 23).

> Where goals were once focused on weight and volumes, Waste Management—along with many other companies, cities, states and even academic institutions—has turned to goals based upon environmental attributes, most notably reduction in greenhouse gas (GHG) emissions through a life cycle approach to assessing recycling.
>
> **(WM 2018, 23)**

9.7.13 Reduce Waste and Consumption of Resources (Materials, Fuel, Energy), and Commit to Recovery and Recycling, as Opposed to Disposal Where Feasible

Commercial and residential carpet tiles, cell phones, refrigerators, VCRs, and CRT glass from computer and television monitors are just a few examples of consumer goods that can be recycled to avoid wasting resources.

Johnson & Johnson's recycling efforts have saved the equivalent of 370,000 trees each year. Since 1972 its energy conservation program has saved the equivalent of 32 million gallons of oil. Since 1988 the company has eliminated more than 5.5 million pounds of packaging from 32 products.

Xerox has implemented programs that enable its customers to return supplies to Xerox for reuse and recycling, such as copy and print cartridges and toner containers.

IBM Research has developed a new process for plastics recycling called VolCat (short for Volatile Catalyst), which turns waste PET into a pure material that could be used to make new plastic material; this process works with materials that are contaminated and otherwise currently not suited for recycling.

Recycling is important; however, elimination is better because corporate and natural resources are spared when the waste is never created in the first place. Baxter is one of the companies that follow this. By 2018, Baxter had eliminated 1,593 metric tons of packaging material using sustainable design as its benchmark (Baxter 2018, 11). To improve their competitive position, not incrementally but radically, forward-looking companies from around the world are embracing the ultimate environmental goal: zero emission (Mcinerney and White 1995).

9.7.14 Management of Environmentally Risky Products

Consistent with the "zero waste equals zero defects" philosophy of Mcinerney and White, Chevron has eliminated products, such as agricultural chemicals and paint thinners, where the costs to properly manage the products' environmental risks and meet regulatory requirements outweighed their earning potential. Monsanto has bioengineered numerous plants to reduce the use of pesticides and to make herbicides more effective and less harmful to the environment. Johnson & Johnson, beginning in 2006, has conducted Environmental Risk Assessments on all of its active pharmaceutical ingredients, and on the majority of its legacy active pharmaceutical ingredients, in order to determine the impact on the environment. In the case of their personal-care products, the lack of an industry-wide standard allowed J&J to develop their own risk assessment methodology for those products. Waste Management probably has the most sophisticated methods of dealing with "hard-to-handle" materials, given the nature of their business and their commitment to "at your door special collection services" (WM 2018, 50).

9.7.15 Education and Training

An environmental policy can include a commitment to communicate appropriate environmental impacts information to all of the business's employees and provide them with the necessary information and skills to deal with the environmental impacts issues and risks that they may encounter. The NCPL Commission noted that some companies develop a legal-risk analysis to help train their employees on communications strategies. This involves determining the business's activities, evaluating and prioritizing the type of legal risks encountered during those activities, and developing an appropriate communications program to manage and minimize risks (NCPL 1996).

In Dow Chemical's 2018 Sustainability Report, the company heralded the launch of its Sustainability Academy, in partnership with the University of Michigan. The Sustainability Academy trained 116 employees in three cohorts, giving them "the hands-on experience and tools needed to bring sustainability business insights into their jobs." Dow is evaluating the program for expansion.

9.7.16 Sustainability Reporting

Sustainability reporting has become a norm in business practice worldwide. This means reporting is indispensable for any company looking to compete in a global market. The following are summaries of the best current sustainability reports:

Apple 2017

Lisa Jackson, former EPA Administrator and current Vice President, Environmental Policy and Social Initiatives, issued Apple's 2019 Environmental Responsibility Report. According to Ms. Jackson,

> Creating powerful solutions to push humanity forward takes relentless innovation. Resolving to do this without taking precious resources from the planet means holding ourselves and our suppliers to even higher standards. We know that accomplishing this work will require all of our best efforts. At Apple, we are committed to building ground breaking products and services with the mission to leave our world better than we found it.
>
> **(Apple 2019, 3)**

Three key components of this system are (1) emissions reductions; (2) component reductions; and (3) 100% renewable energy (Apple 2019, 7).

Evonik Industries 2017 (Germany)

Evonik provides a clear outline of its environmental, safety, health, supply chain, and employment equality targets for the coming years. Evonik's incorporation of the United Nation's 17 Sustainable Development Goals in its report allows the company to highlight how each company target is working toward sustainable practices on a local, regional, and global level. The UN states that "[t]he Sustainable Development goals are the blueprint to achieve a better and more sustainable future for all. They address the global challenges we face, including those related to poverty, inequality, climate, environmental degradation, prosperity, and peace and justice."

Hongkong and Shanghai Hotels 2017 (China)

Hongkong and Shanghai Hotels provides a report detailing sustainability benchmarks within their manifold operations. This includes development involving employees, supply chain, operations, surrounding communities, and guest experience. The report also highlights how partnering with non-profits has allowed them to expedite sustainability initiatives.

CH2M (now Jacobs)

CH2M's "Sustainability and Corporate Citizenship Report" from 2017 developed "key performance indicators" and then performed a comparison to business goals. Since 2005, CH2M was ISO 14001 compliant in the U.S. and Canada and was a forerunner among U.S. engineering companies in developing environmental management systems, before merging with Jacobs in 2018. Jacobs' 2018–20 sustainability strategy introduces "Plan Beyond: Sustainability Today for Tomorrow." Chair and CEO Steve Demetriou writes, "For us, this means social and economic progress while protecting our environment and improving resilience. It is also about being the employer of choice for our people and being the go-to solutions provider for a more connected, sustainable world" (Jacobs 2019, 3).

Entel 2017 (Chile)

Entel, Chile's largest telecommunication provider, faces unique environmental performance challenges in the developing world. Entel outlines its operations, which display a commitment to improving social welfare through increased connectivity yet not at the expense of sustainable expansion practices. The Entel report details its involvement with UNESCO and other international groups that seek to ensure responsible corporate expansion in developing nations.

Xcel Energy

Xcel's "Destination 2050" Corporate Responsibility Report from its Chairman, President, and CEO Ben Fowke states,

> At the end of 2018, we became the first energy provider in the country to announce plans to serve customers with 100% carbon-free electricity by 2050. The theme of this report—Destination 2050 Building the Future—describes our bold vision for a carbon-free future while delivering the reliability and affordability that our customers need and expect.
>
> **(Xcel Energy 2019, 6)**

BP 2018

BP's 2018 Sustainability Report is titled "Responding to the Dual Challenge." "As the world demands more energy to fuel increasing prosperity and provide people with a better quality of life," it states, "it also demands energy delivered in new ways, with fewer emissions" (BP 2018, 1). BP's sustainability analysis, running 80 pages, details its plan to transition to low carbon emissions; reduce greenhouse gases; create energy efficiency in BP's products; enhance future mobility and electric vehicles; create carbon capture, use, and storage; and increase renewable energy (BP 2018, 5).

ConocoPhillips 2018

This company's "2018 Sustainability Report" is a 207-page strategic analysis setting forth an immensely complicated risk-based calculation of the company's transition to low-carbon fuel usage. The emphasis is continuously on lowering GHG emissions, with an annual reduction of 7 million tons since 2009 (ConocoPhillips 2018, 85), and carbon capture. Its use and sequestration statistics culminated in 2018 with the company selling 530,000 tons of CO_2 from process emissions to a third party.

Chevron 2018

Not to be outdone by ConocoPhillips, Chevron published its 2018 Corporate Responsibility Report, with highlights including its support of the 17 United Nations Sustainable Development Goals and declaring that "the Chevron Way" would emphasize (1) protection of the environment; (2) valuing of diversity and inclusion; (3) advocacy of governance by getting the right result the right way; and (4) performance. Both ConocoPhillips and Chevron tackle their strategy as to how two giant oil and gas companies can respond to and "manage" climate change.

All of these reports share similar components and goals that allow for their success. Every company displays a highly developed information technology system. These enable companies to identify their annual benchmarks and set realistic improvement goals. Further, each company displays a commitment to improved environmental performance that begins with top management and is instilled in employees of every level. Last, these companies hold direct partners as well as themselves accountable for environmental progress. These factors, combined with market-relevant information, allow their reports and therefore their company as a whole to stand out as leaders in improving environmental performance.

9.7.17 Encourage the Use of an Environmental Management System by Suppliers and Contractors

Baxter's environmental policy includes a commitment to work with its suppliers and contractors to enhance environmental performance. Bristol-Myers Squibb goes further by stating that when feasible, it will give preference to suppliers and contractors whose environmental health and safety commitment and practices are consistent with its own and who have demonstrated environmentally responsible products, services, and management. Apple has requested its suppliers to commit to power all of their Apple production with 100% renewable energy. According to its 2019 Progress Report, Apple states that 44 Apple suppliers have committed to these goals, making Apple on track to far exceed its 2020 goal of bringing 4 gigawatts of new clean energy into its supply chain.

There are compelling environmental and legal reasons for this management commitment to working closely with suppliers to achieve these goals. Back in 1996, the NCPL Commission said that to ensure accountability mechanisms apply to all sources of liability and legal risk, a company may wish to extend some aspects of its management system practices outside of its organization. For example, some actions of external corporate agents or independent contractors can create significant liability for the company. Accordingly, a business has a strong interest in holding these individuals accountable for their environmental impacts. Companies may wish to incorporate provisions in related contracts requiring conformance by agents and contractors acting for the firms and providing for reporting and reviews concerning conformance by these outside parties.

The Commission also found that a company may wish to inform outside parties with which it conducts business of the company's environmental expectations. For example, a company may want to consider requiring specific compliance results in agreements or contracts with agents and vendors. As a part of its environmental management system, a company may require that its agents and vendors be ISO 14001 certified. This requirement can be included in the written contracts that the company negotiates and may also be a part of a vendor certification process.

While the environmental policy is perhaps the centerpiece of an environmental management system, it is still only one component of a fully integrated plan to minimize environmental risks and reduce impacts. In our first edition, we identified other elements of compliance that round out an effective environmental management system.

9.7.18 Environmental Aspects

Most formal environmental management systems recognize that a business's policy, objectives, and targets should be based on knowledge about environmental aspects and significant environmental impacts associated with its activities, products, or services. This can ensure that significant environmental impacts associated with these aspects are taken into account in setting environmental objectives. In order to manage and reduce environmental risks and liabilities, a company needs to know the environmental aspects of its activities, products, and services.

The most obvious environmental impact of a company is its waste streams. Elimination of waste streams directly cuts down on environmental aspects and impacts. In the 2018–19 environmental-sustainability reports, multiple companies discuss ways they plan to eliminate waste streams, such as Apple's zero waste (Apple 2019, 4) and HP's zero deforestation (HP 2018, 2A). As mentioned, Waste Management's Life Cycle Assessment Approach to Recycling focuses on environmental attributes, most notably the reduction of GHGs (WM 2018, 23). Other impacts may be minimized by factors such as how products are designed and packaged, how energy sources are used, and how products are manufactured—to name a few. Companies can have energy audits conducted to eliminate unnecessary uses of energy and natural resources. In 2018 alone, Graham Partners' portfolio companies realized $2.1 million in annualized savings from implementing energy, water, and waste reduction programs. Proactive companies need to establish and maintain procedures to evaluate the impacts of their operations on the environment.

Some environmental impacts are not so obvious. But unless they are identified, they can cause liability for the company. The NCPL Commission found that in order to contain risks, companies need to identify nonobvious and incipient misconduct that tends to promote illegal actions. For example, if a company has a waste disposal policy that requires employees to use haulers that charge the lowest disposal costs possible, illegal discharge of the company's waste may result. Similarly, if corporate environmental goals are unrealistic, employees may be tempted to lie in reports in order to appear to achieve those goals. A company has to be prepared for a full range of collateral, unwanted consequences if it sets the pollution bar too high.

Changes in operations, products, or services may affect the environmental aspects and their associated impacts. A management system must be responsive to daily operations and be dynamic to account for changes in business activities. The system must be appreciated by those faced with new and challenging environmental issues. Employees learn something new every day and face uncertainty in solving complex problems presented by the workplace. Responsive environmental management systems have built-in components to instill confidence in employees and to motivate them in the face of uncertainty. One company explained its program to employees by illustrating its intended implementation with concrete examples drawn from the business's specific work-related activities. These illustrations can take into consideration changes that can occur in the workplace as well as failures and how to responsibly deal with them.

Before a process failure occurs, companies should have plans in place to deal with the consequences both internally and externally. DuPont agreed to inform its employees and the public about the safety and health effects of its products and workplace chemicals and to provide leadership in establishing programs to respond to emergencies involving hazardous materials in communities where the company has a significant presence. Conoco agreed to maintain rigorous emergency preparedness plans and response capabilities.

In addition to developing hazard awareness and crisis management plans, companies need to find out how frequently a situation may arise that could

lead to the impact. This will determine the level of effort required to minimize the risk of occurrence of an incident.

To understand the degree of risk, most formal EMSs require the following
environmental concerns to be addressed: the scale of the impact, the severity
of the impact, the probability of occurrence, and the duration of impact. Once
these concerns have been considered, formal EMSs typically require that the
following business concerns be addressed as well: the potential regulatory
and legal exposure, the difficulty of changing the impact, the cost of changing
the impact, the effect of change on other activities and processes, the concerns
of interested parties, and the effect on the public image of the organization.

Environmental management systems will be most useful when they are
carefully tailored to fit each business's unique situation. The same may be
true regarding operating units within a business. Each different operating
unit of a business may need its own environmental management system. The
system must be adapted to address multiple different and possibly inconsistent legal requirements. For example, many companies with global operations must confront numerous different, and potentially conflicting, legal
and environmental requirements. Several international companies adopt a
core of global standards but permit local management to modify portions of
their program to take into account local needs and requirements.

9.7.19 Legal and Other Requirements

A formal procedure is required to identify ongoing legal and other requirements
that are applicable to the environmental aspects of the business. The standard
requires identification of not only governmental regulations, but also industry
associations or groups and commercial standards of practice and professional
codes. Under the systems approach to the minimization of risk, the procedure
should also identify the range of possible consequences of a business's actions,
including civil and regulatory violations. Internal controls procedures can then
be implemented to effectively eliminate noncompliance. The identified requirements must be readily accessible, updated, and available to employees.

For compliance purposes a business needs to identify regulatory requirements that are specific to its activities (e.g., site operating permits), to its
products or services, and to its industry. The business also needs to identify
relevant general environmental laws, authorizations, licenses, and permits.
To keep track of all relevant legal requirements, a business can establish and
maintain a list of all laws and regulations pertaining to its activities, products, or services in electronic or hard-copy format and a procedure for identifying changes that occur to legal and other regulatory requirements.

In creating a compliance program, a business should thoroughly examine its liability-risk profile. To accomplish this profiling, it may be useful to
examine the risk experience of the business's industry. For example, a company can have its legal department prepare a report on compliance problems
that the company's industry has experienced. To complete a liability-risk

profile, a company can assess its own past compliance history. One company assigned an in-house attorney to conduct both a search of the company's files and a compliance audit to determine compliance risks that the company had faced. It should be noted that while the involvement of an attorney may help preserve the attorney-client privilege, to the extent that such studies are viewed as management tools or are disclosed to public agencies to demonstrate compliance diligence, these audits can be subject to disclosure in civil and criminal cases.

Management systems should help companies focus upon risks that a business most frequently confronts. A listing of recently encountered environmental risks and problems can be useful in ensuring that a complete set of corresponding compliance program elements is adopted. Companies that operate within a broad regulatory framework can begin their risk assessment by examining their own histories of violations and citations and those of other similar companies. Many companies also conduct "litigation audits" as a starting point for assessing their legal risks.

9.7.20 Objectives and Targets

Most formal EMSs require that there be documented environmental objectives and targets set at each relevant function and level within the business. Companies need to set forth these goals and methods for achieving them in a clear and straightforward manner. Objectives are the overall goals for environmental performance that are identified in the policy statement. When establishing its objectives, a business should take into account the relevant findings from the initial review and the identified environmental aspects and associated impacts. The targets should be time-specific and measurable. When the objectives and targets are set, the organization should consider establishing measurable environmental performance indicators. Objectives and targets can apply broadly across an organization or more narrowly to site-specific or individual activities.

The environmental objectives and targets need to reflect both the goals of the environmental policy and significant environmental impacts associated with the business, activities, products, or services. In setting objectives and targets, management has to consider whether the employees responsible for achieving the objectives and targets have had sufficient input into their development and whether the views of interested parties have been adequately considered.

The following are some examples of companies who have made public commitments to achieving environmental objectives and targets.

Reduce waste and the depletion of resources

In 1991 Hitachi was one of the first Japanese companies to set industrial waste reduction goals. Since that time virtually all large companies provide waste reduction objectives and targets. Hitachi in its 2018 Sustainability Report, with the tagline "Inspire the Next," has stuck to these themes aiming for a low-carbon, resource efficient, and harmonized society (Hitachi 2018, 91). Hitachi's specific goals included (1) 80% reduction of value chain CO_2 emissions; (2) 50% improvement in water use efficiencies; and (3) minimizing impact on natural capital (Hitachi 2018, 92). As of 2018, Hitachi reports that it has installed 700 water purification plants and 900 sewage treatment plants in Japan (Hitachi 2018, 98). In its 2018 Environmental Responsibility Report, Apple stated these targets on climate change: 100% of global facilities powered by 100% renewable energy; 70% decrease in average product energy use in 10 years; and 25% reduction in overall carbon footprint compared to 2015. It listed these targets for resources: 100% recycled aluminum enclosures in MacBook and Mac Mini; 100% responsibly sourced wood fiber in all retail packaging; and 145 priority materials for transitioning to 100% recycled or renewable content (Apple 2018, 5).

Design products to minimize their environmental impact in production, use, and disposal

Many companies, like Apple, have committed to designing products for disassembly so that they can be recycled into new products. Apple's Daisy disassembly robot takes apart used products for reuse. Takeback laws in Japan and Germany have provided financial and ecological incentives to companies to redesign products ranging from vacuum cleaners to sports cars to reduce environmental impacts. As you likely know from the label on your computer monitor, the EPA has signed partnership agreements with office equipment manufacturers representing approximately 90% of the office equipment market to produce computers, monitors, printers, and facsimile machines that automatically power down when not being used.

Control the environmental impact of sources of raw material

WMX Technologies has committed to using renewable natural resources, such as water, soils, and forests, in a sustainable manner and to offering services to make degraded resources once again usable. WMX has also agreed to conserve nonrenewable natural resources through efficient use and careful planning.

Minimize any significant adverse environmental impact of new developments

Chevron has committed to work closely with local tribes and government officials in New Guinea to ensure that the cultural and environmental integrity of the region is retained. Chevron has buried pipelines to minimize rain forest damage and helped to build schools and clinics.

Sun Company has committed to pay special attention to the protection of the surrounding environment at present facilities and when planning for new facilities or operations.

In the 1990s, Westinghouse had committed to minimizing adverse environmental impacts in the planning and development stage of new infrastructure. Westinghouse's 2018 EHS and Sustainability Policy states that it is "committed to ... protecting the environment by minimizing the raw materials and energy usage while reducing waste, preventing pollution, and re-using and recycling materials and resources to the extent that is economically and technically feasible" (Westinghouse 2018, 1).

Promote environmental awareness among employees and the community

Chevron's Citizen's Advisory Group at its Oak Point Chemical Plant is composed of civic leaders and concerned members of the community who are reviewing the company's plans to handle emergency incidents. Chevron also conducts and periodically updates training programs to help employees understand how social, political, and legal aspects of society's environmental values affect its business. The training program emphasizes the company's dedication to environmental compliance and risk management.

In our first edition, we discussed how Howe Sound Pulp and Paper Corporation held public meetings in Canada with slide shows and videos to show its commitment to minimize environmental impacts in Canada's forestry industry. By 2018, this program also included (1) generating green power from renewable resources; (2) safety management for its employees; (3) work place productivity; and (4) environmental stewardship.

The management of Inter-Continental Hotels (IHG) held meetings with its employees to demonstrate its commitment to reduce its environmental impacts. Subsequent meetings produced tangible benefits by reducing impacts and greening the hotel system. By 2018, IHG in its Responsible Business Report stated that it reduced its carbon footprint by 2.2% per occupied room that year and launched two water projects in London and Delhi (IHG 2018, 33).

Environmental performance indicators can be used to measure progress toward an objective

Examples of performance indicators include the quantity of raw material or energy used and the quantity of emission of gases such as CO_2.

As noted in our first edition, Nissan recognized that the long-term viability of its business may depend on developing alternatives to the internal combustion engine. Nissan, like most automotive companies, has invested heavily in its ability to make a significant contribution to the reduction of conventional pollutants such as CO_2. In its most recent report, Nissan stated that it is taking steps to (1) identify direct and indirect environmental efforts of its activities; (2) promote human kindness and concern to form the basis of environmental protection; and (3) move to zero emissions through the elimination of CO_2 emissions from new Nissan vehicles.

Many companies like Sun Company are measuring the reduction in the use of water in its facilities in an effort to decrease the quantity of its water discharge streams. Companies that filed sustainability reports in 2018–19 with lengthy explanations concerning their water usage include Apple, HP, IBM, Jacobs, Lockheed Martin, and Waste Management.

Monsanto Company pledged to reduce all toxic and hazardous releases and emissions, working toward an ultimate goal of zero effect. That was in the 1990s. Bayer acquired Monsanto and in its 2018 Annual Report emphasized the development of management systems including sustainable management to safeguard societal and economic viability integrating "sustainability into our work procedures" (Bayer 2018, 1.2. Strategy and Management, 5).

Percentage of material recycled and used in packaging

In the 1990s, Coors measured the amount of waste it has recycled, including 107 tons of office and colored paper; 7,703 tons of corrugated paperboard; 19,568 wooden pallets repaired; 16,965 cubic yards of wooden pallets composted; 1,087,600 gallons of process sludge composted; 125 tons of stretch wrap; and 89,000 pounds of plastic banding. Following its merger with Molson, the combined company set six goals for 2025 including reductions in CO_2 emissions, protection of water resources and water use efficiency, and zero waste to landfills, which has been achieved at 14 facilities (Molson Coors 2018, 5).

In the 1990s, Johnson & Johnson calculated that its U.S. paper recycling efforts alone are saving the equivalent of 370,000 trees each year. In its Sustainability Report for 2018, J&J's approach is optimizing operations, reducing life-cycle impacts, encouraging suppliers to make environmental improvements, and partnering with stakeholders (Johnson & Johnson 2018, 12).

Union Carbide's efforts to collect paper, plastic, metal, and used oil at U.S. sites yielded 17 million pounds of recyclables in 1992. The 1993 total was 19 million pounds. Paper recycling, including cardboard and newsprint, totaled 1.5 million pounds in 1992. The 1993 total was 2 million pounds. These efforts saved 17,000 50-foot trees; 7 million gallons of water; 2,500 barrels of oil; and 3,000 cubic yards of landfill space.

Xerox uses permanent parts in its shipping containers and carts in manufacturing operations; these were returned to suppliers for refilling and reuse. Xerox also uses recycled-content corrugated cartons for shipping consumable supplies. In its most recent report, Xerox states that it has achieved a 35% water reduction from 2010 level and reused 300,000 of reject water in cooling towers. It is also identifying sources of air emissions and tracking them.

9.7.21 Environmental Management Program

Sustainability reporting needs to include specifics regarding how environmental management programs establish achievable and identifiable objectives and targets. A program is necessary to achieve good environmental results. The program is only a part of the larger environmental management system designed to manage and reduce environmental risks, impacts, and liabilities. It must define the means, schedule, and responsibility at each relevant function and level to achieve the objectives and targets. It should also be amended as necessary to include new developments or modified activities and products or services, yet still attain the objectives and targets.

A written report of how the environmental management program was created and implemented can be useful. The program needs to have sufficient detail for an employee to clearly understand how it was first developed and whether the planning process involved all responsible parties. The environmental management program should document the specific steps necessary for the business to succeed in achieving objectives and targets and who is responsible for achieving those goals.

A written record can also be prepared to demonstrate how the system operates on a daily basis and to monitor the system's current effectiveness. A written record can also be used to defend the program's effectiveness and tailor the program for more effective use in the future. The environmental management program also should have a process built in for periodic reviews of the program.

An effective environmental management system, which includes a compliance component, is a dynamic process that can be designed to be flexible and modified, when appropriate, to reflect changed circumstances. It must be able to respond to new business activities or other significant changes in the business. A successful program is usually adaptive to changes in the company's environment and can include components that respond to the new plans and conditions of the business. Unplanned changes can produce new environmental and compliance risks, so an effective program should be able to adapt to meet those risks without compromising the system's integrity.

The following are some examples of how businesses maintain their programs in the face of changing circumstances.

Conoco's environmental program dates back to February 29, 1968. To adjust the program to changing conditions, Conoco regularly conducts environmental quality assurance audits as one way to determine if it is achieving its goal of conducting business with "respect and care for the environments in which [Conoco] operates." As mentioned, ConocoPhillips' 207-page "2018 Sustainability Report" is the most comprehensive analysis of its transformation to "low carbon fuel usage." This company has been tracking its environmental impacts for decades and trying to reduce those impacts for future generations.

Sun Company, Inc. changed the direction of its environmental program in 1993 when it became the first Fortune 500 company to endorse the CERES Principles in order to demonstrate its commitment to public accountability for environmental protection. Since then, Sun has published an annual report on its performance and accomplishments of its health, environmental, and safety program and reports on its progress in consistent, measurable terms.

In our first edition, we noted IBM conducted audits and self-assessments of its compliance with its environmental policy, measured progress of its environmental affairs performance, and reported periodically to its Board of Directors. By 2018, in its Corporate Responsibility Report, IBM had published a 54-page summary of how "Trust and responsibility [are] earned and practiced every day," according to Chairman, President, and CEO Ginni Rometty. One purpose of the report is to highlight how IBM procures 37.7% of its electricity from renewable sources and has a goal of 55% by 2025. In addition, it has developed a "Volcat" process that is designed to turn used plastic bottles into material that can be used to manufacture new plastic products. More importantly, the report lays out in detail how the environmental program at IBM works. This report narrates IBM's commitment to environmental leadership and adds a significant "social impact" of its operations with the aim to build stronger communities worldwide, focusing on education, health, and disaster resiliency (IBM 2018, 38).

As described in our first edition, Johnson & Johnson instituted a periodic environmental assessment program conducted by independent outside reviewers. More than 200 of its worldwide facilities are covered by these third-party audits. When no standards are specified, or in countries where environmental standards are less stringent than in the U.S., Johnson & Johnson followed its higher corporate requirements and guidelines. In its most recent report, Johnson & Johnson's approach is "based on optimizing operations, reducing lifecycle impacts, encouraging suppliers to make environmental improvements and partnering with stakeholders" (Johnson & Johnson 2018, 12).

9.7.22 Structure and Responsibility

Most formal EMSs require environmental roles, responsibilities, and authorities to be defined, documented, and communicated to all relevant employees. Financial, technological, and human resources essential to the environmental management system must also be provided. A management representative must be selected who reports directly to senior management on environmental management system performance and improvement. The management representative must ensure that the environmental management system is established, implemented, and maintained.

It is important for the individual who is selected as management representative to take ownership of the environmental management system and to exert overall responsibility for initiating, coordinating, and monitoring the environmental management system efforts. This individual obviously has to have the necessary degree of clout and respect within the business to make the system work. A high level of authority, coupled with access to senior management, assures that the management representative and the environmental management system are perceived as important activities by the other people in the business. The individual must also have the right combination of personal characteristics and knowledge of systems operations to be effective in leading and promoting environmental performance improvements. Businesses with regulatory responsibility need to have a carefully defined program that weeds out managers and employees who disregard the law, regardless of the consequences to business operations.

At HP, for example, the "Global Head of Sustainability and Product Compliance" has her performance and compensation directly associated with the management of sustainable impact and the achievement of related targets and metrics, both public and internal, including product energy efficiency. Other executives have quarterly goals/metrics tied to the company's sustainable impact strategy—including GHG emissions reduction goals. Several other HP directors and manages have a component of their total compensation (salary and bonus) based on responsibility for, and effective implementation of, corporate initiatives to address climate change (HP 2018, 5).

At Lockheed Martin, the

> formal sustainability structure is made up of our Board of Directors, executive leadership team and key functional leaders responsible for sustainability initiatives. Our lead sustainability executive is the Senior Vice President (SVP) Ethics and Enterprise Assurance (EEA) who oversees ethics; enterprise risk; environment; safety and health; internal audit; and sustainability … . Incentive compensation for Lockheed Martin executive is linked to sustainability factors we measure and report.
>
> **(Lockheed Martin 2018, 11)**

Large businesses need to be careful to allocate sufficient resources to meet the goals of the environmental management system. The goals are broader than merely the objectives and targets of the environmental management program. They include the need for the business to minimize environmental risks and liabilities, as well as other goals such as promoting effective communications, training, and improving industry standards. The management representative is responsible for ensuring that the company finds the correct level of financial contribution to the system to achieve these goals.

Businesses with limited resources can team with larger companies to share technology and compliance information. They can also create opportunities for similarly situated companies that are interested in developing environmental management systems to define and address common issues, to share knowledge, to facilitate technical development, and to use facilities and consultants jointly to develop environmental management systems. To save expenses, smaller businesses can participate in training and awareness programs conducted by standardization organizations, consultants, associations, and chambers of commerce, as well as universities and other environmental research centers.

In order to have a successful environmental management system, participation in and responsibility for the system should not be limited to the management representative. Environmental responsibility needs to be shared throughout the business. Good examples of a comprehensive oversight structure are in Jacobs Engineering (Jacobs 2019, 11), Nissan's governance policies and philosophy (Nissan 2018, 219) and Waste Management's comprehensive compliance controls over its waste intensive business (WM 2018, 168–75). Senior management in these large companies ensures that incentives for making the system work are uniform, consistent, and spread throughout the business. Incentives and disincentives are important tools for promoting awareness of environmental responsibility and achieving the goals of the environmental management system.

The NCPL Commission found that corporate managers can influence behavior by linking employment treatment with each employee's environmental performance—for example, linking increased compensation and advancement to employees' furtherance of compliance goals and objectives. Employees throughout the business can be told that it is the policy to allocate incentives and disincentives in accordance with each employee's environmental performance. Corporate leaders can also underscore this message to ensure that the rewards and discipline are in accordance with the relative levels of effort (NCPL 1996).

Environmental responsibility includes the development of a risk reduction litigation management plan, which can include litigation awareness programs and other techniques to avoid or at least minimize liability and legal costs. The management representative can coordinate the development of such a strategy with legal counsel who specializes in risk reduction techniques and identification of liability-causing conduct. The plan should identify the company's overall goals and objectives for the litigation and create a cost-benefit analysis. The plan should address concerns about adverse publicity, as well as negative impacts and possible benefits of litigation. The plan should also include steps the company needs to take if the government commences an investigation; who should retain experienced in-house or outside counsel; how, when, and who should be responsible for analyzing key facts; and who should retain experts and when should they be retained. Companies need to know their corporate history and insurance profile to be able to respond quickly and efficiently to any type of environmental claim whether it comes from the government or a third party, and to be able to notify carriers promptly. Insurance needs have to be addressed on a continuing basis and analyzed according to risk history.

The management representative should have a plan in place to determine how any type of case should be staffed and handled, either internally or using outside counsel when appropriate. The management representative can also be in charge of developing opportunities for employees to learn alternative dispute resolution methods and techniques to promote understanding that such techniques can be used to minimize litigation costs and provide better resolution of disputes.

A risk reduction-litigation management plan should include the following questions in determining whether ADR may be appropriate to resolve the dispute:

- Are there present or foreseeable difficulties in the negotiation that will require time or resources to overcome in order to reach settlement?
- Is the case negotiable, i.e., no precedent-setting issues are involved?
- Is there enough case information to substantiate the violations or claims?
- Is there sufficient time to negotiate in light of court or statutory deadlines, or are the parties to sign a tolling agreement (an understanding that a statutory deadline for starting a lawsuit will be extended)?
- Are all the stakeholders identified? Can they all be persuaded to come to the table?
- Is sufficient authority present? Is each relevant interest group adequately represented?
- Is the table level? Can it be made level?
- Is there a climate of trust and willingness to negotiate—or a way to get there?
- Do the parties agree on the scope of issues to be negotiated or are they willing to allocate time and other resources to coming to agreement?
- Will it be possible to bring the best available information and expertise to the process?
- Can fundamentally different values and assumptions be identified and discussed?
- Will the group be able to identify a sufficient number of legal and economically feasible solutions?
- Are the parties committed to the process?
- Are they prepared to analyze costs and benefits through joint problem solving?
- Are they prepared to sign a written agreement?
- Are they capable of implementing possible solutions?

There is no one right way, no one right process, for efficient and effective ADR. Flexibility in approach, and familiarity with a range of dispute management tools and approaches, will permit the management representative to most effectively manage the company's environmental disputes.

9.7.23 Training, Awareness, and Competence

It is recommended that a conformance training and awareness program be instituted to identify training needs and to ensure that each employee whose work has the potential to create a significant environmental impact receives the appropriate training. Training procedures can be instituted to make employees aware of the importance of conformance to the environmental policy; procedures and the requirements of the environmental management system; significant environmental impacts of their work activities; environmental benefits of improved personal performance; roles and responsibilities in achieving conformance to the policy, procedures, and requirements of the environmental management system; and the potential consequences of departure from specified operating procedures. See for example, Waste Management's training protocols (WM 2018, 171); Xcel's training program (Xcel 2019, 96–7); Nissan's training and education of employees (Nissan 2018, 118); and HP's training inside and outside the company (HP 2018, 39). IBM states it has invested in more than 24 million hours of professional education for IBMers. Many of these hours are devoted to sustainable practice of the company.

Employees should be trained at all levels to impress on them the importance of disclosure of possible violations, as well as the procedures to do so. Legal counsel can assist companies with the development of litigation awareness programs to identify liability-causing conduct. Typical training programs include a variety of topics related to legal requirements, company values, and the means to consider these in company business decisions and actions (NCPL 1996). Environmental training goes beyond aspects of health and safety by requiring proven competency. Personnel whose tasks can cause significant environmental impacts must be competent based on education, training, and experience. Proper training will require employees to be responsible and accountable for their business's environmental impacts.

Education and training programs are important to ensure that employees have appropriate and current knowledge of regulatory requirements, internal standards, and the business's policies and objectives. The level and detail of training, however, may vary according to the tasks performed by employees. It is important to match training to the tasks routinely handled by the employees.

Training programs need to be designed so that employee training needs are identified and plans developed to address their specific needs. The NCPL Commission suggests that an effective training program can be targeted to reach an intended audience and should be understandable, accessible, and practical (NCPL 1996). Interactive computer training programs and software packages are commercially available that have the capability of providing many employees with effective training. The employees' performance is monitored and results tabulated for review and evaluation purposes. Formal documentation of the type of training given and the results of the training exercises is useful to keep a record of compliance activities and to measure the effectiveness of the training received.

9.7.24 Communications

As previously discussed, communications can include establishing processes to report internally and externally on the environmental activities and issues to demonstrate management's commitment to improving environmental performance. The company also has to deal with concerns and questions raised about the environmental aspects of the business's activities, products, or services. At the same time, communications need to raise awareness among stakeholders of the business's environmental policies, objectives, targets, and programs.

The NCPL Commission found that a good business practice is to have a communications component in the compliance program that makes employees and other agents aware of the applicable standards of conduct and to promote compliance (NCPL 1996). The formality of the communications component is likely to increase with the size of the business.

Companies should set up and maintain procedures for internal communications between the various levels and functions of the business. A process needs to be created for receiving, documenting, and responding to employees' and other interested parties' concerns.

The business also needs to have a documented plan for communicating the business's environmental policy and performance. A plan to disseminate the results of environmental management system audits and reviews to appropriate people in the business is also necessary. Protection of the audit results from unauthorized personnel and third parties is important, along with keeping the results confidential and privileged where appropriate.

Public environmental reporting is becoming a popular way for companies to distribute their environmental results. Environmental reports, as discussed in earlier chapters, can include the business profile and the text of the environmental policy, objectives, and targets. Beyond this, ISO 14004 recommends that the public report also include environmental management processes (including interested-party involvement and employee recognition), environmental performance evaluations (including releases, resource conservation, compliance, product stewardship, and risk), opportunities for improvement, supplementary information (including glossaries), and independent verification of the contents. The level of detail contained in these public reports should be directly related to the degree of exposure management wants to give its environmental management system. Companies with relatively new systems should examine this question carefully and consider alternatives to public reporting until the effectiveness of the company's program is proven and the company is ready to discuss openly its environmental record and future plans.

Internal and external environmental communications and reporting involve two-way communications that require the dissemination of understandable and adequately explained information. Stakeholders should receive accurate and verifiable information pertaining to environmental performance and be given a meaningful opportunity to respond to the information and make constructive suggestions for improvements.

The management representative and other individuals involved in the development and operation of the environmental management system need to have training on collaborative decision-making techniques in order to learn how to reach consensus with internal and external stakeholders. This process provides an open forum for all stakeholders to be proactively involved in reaching collaborative decisions while reducing the likelihood of internal conflicts with employees and litigation with external stakeholders.

9.7.25 Environmental Management System Documentation

Business needs to establish in paper or electronic form a description of the core elements of the environmental management system and their interaction, together with directions to related documentation. Documentation of the system itself will vary according to the nature and size of the business. The documents are the most important vehicle to promote employee awareness of what is required by the system, to achieve the business's environmental targets and objectives, and to enable the evaluation of the system and environmental performance. ISO 14004 recognizes that where elements of the environmental management system are integrated with a business's overall operations, the environmental documentation should be integrated into existing documentation. The environmental management system documentation should include the environmental policy, objectives, and targets; the means by which environmental objectives and targets are achieved; a description of the key roles, responsibilities, and procedures; the location of and means to retrieve documentation; and other relevant elements of the business's environmental management system. The documentation can also discuss how the environmental management system elements are to be implemented.

The business needs to have a process set forth in the environmental management system for developing and maintaining documentation. The documentation system should be integrated with existing documentation and be available to all employees who have responsibility for maintaining records. Documentation must be managed so that the documents are readily identifiable, organized, and retained for a specified period using codes for the appropriate organization, division, function, activity, and contact person.

Documents should be periodically reviewed, revised as necessary, and approved by authorized personnel prior to issue. The current versions of relevant documents need to be available at all locations where operations essential to the effective functioning of the system are performed.

Procedures need to be designed for the control of relevant documents for the environmental management system. The location of controlled documents must be established and known by the management representative and designated personnel. The document control system will require a periodic review, and any necessary revisions should be made with authorized approval for adequacy. When documents are created, document control procedures need to be carefully implemented so that the business can maintain its compliance record in an organized and retrievable manner and prevent destruction of relevant documents that can be used affirmatively to prevent a claim from being filed. Good document control ensures that the company demonstrates evidence of compliance and a commitment to environmental excellence. Current versions of relevant documents must be available where essential operations are performed. Obsolete documents should be promptly removed to prevent unintended use. Obsolete documents that are retained for legal or historical purposes should be suitably identified. The standard requires that documentation be legible, dated, identifiable, orderly, and retained for a specified time period.

9.7.26 Operational Control

Businesses need to identify those activities and operations that are associated with potential environmental impacts and that fall within the scope of the policy, objectives, and targets. These activities, including maintenance, must be carried out under controlled conditions. Businesses need to be careful not to delegate important responsibilities to individuals who are incapable of handling demanding tasks that can lead to environmental accidents or significant impacts. The NCPL Commission found that companies need to exercise due diligence to prevent the delegation of substantial discretionary authority to persons that have a propensity to engage in illegal activities. Placing an employee in a job without sufficient oversight and supervision that results in inadequate performance can lead to criminal or civil violations. In reviewing operational control issues, companies need to be particularly attuned to assigning responsible employees to sensitive positions and to give them sufficient supervision and control to achieve the goals, targets, and objectives of the environmental management system without putting the company in any unnecessary risk.

Operational control includes documenting all procedures for the handling of goods and the providing of services that have the potential to lead to significant environmental impacts. Management has to create documented procedures for any situation where the absence of procedures could lead to procedural deviations or environmental impacts. These procedures must be incorporated into the environmental management system and communicated to employees, suppliers, and contractors.

9.7.27 Emergency Preparedness and Response

Emergency plans and procedures are needed to ensure that there will be an appropriate response to unexpected or accidental incidents. DuPont, for example, supports the chemical industry's Responsible Care® program and the oil industry's Strategies for Today's Environmental Partnership, which are key programs to avoid adversely impacting the environment. DuPont has committed to be prepared for emergencies and to provide leadership to assist local communities to improve their emergency preparedness. Conoco has similarly developed "emergency preparedness plans and response capabilities."

DuPont and Conoco are positive examples of how companies need to create procedures to identify the potential for accidents and emergency situations before they happen. Companies should create procedures regarding how to respond to unplanned events, and how to prevent or mitigate the environmental impacts arising from those occurrences. The procedures need to be documented, reviewed, and revised as necessary, particularly after near misses, accidents, or emergency situations. The procedures should be periodically tested where practical. Kaiser-Hill, for example, at the former Rocky Flats nuclear weapons facility that has stored plutonium, regularly ran mock raids on the facility to test its emergency preparedness to safeguard the plutonium that could be used for terrorist activities.

Emergency plans can document emergency organization and responsibilities; a list of key personnel; details of emergency services (e.g., fire department and spill cleanup services); internal and external communication plans; actions taken in the event of different types of emergencies; information on hazardous materials, such as material safety datasheets (including each material's potential impact on the environment and measures to be taken in the event of accidental release); and training plans and testing for effectiveness.

Consideration should be given to the EPA audit policy and the U.S. Department of Justice's Disclosure Guidelines as to whether to self-disclose. Businesses that have a strong environmental management system, and that make a timely, complete, and voluntary disclosure, have the best opportunity to secure a declination or a civil, rather than criminal, penalty. Before disclosing to the government, a business should correct the violation as soon as possible or create a corrective plan, determine the legal duty to disclose, get an objective opinion of knowledgeable environmental professionals to discuss how to take immediate corrective actions and whether to disclose, consider all related and unrelated collateral consequences that may occur as a result of disclosure, and consider how to eliminate the cause of the violation by implementing an environmental system or changing the existing one to prevent recurrence of violations.

Companies that have crisis management plans, like DuPont, Conoco, and Kaiser-Hill, have taken preventive steps to avoid accidents and those types of catastrophic events that have led to strong public criticism of particular industries in the last several decades. Preventive measures can include regular maintenance and review and redesign of products, processes, and services to eliminate risk-creating circumstances. Professional risk managers and consultants can review operations and make suggestions on how to avoid activities that could be construed as possibly creating opportunities for employees to engage in environmental crime.

9.7.28 Monitoring and Measuring

A good EMS requires a business to monitor and measure, on a regular basis, key operations, characteristics, and activities that may have significant environmental impacts. For example, Apple states, "We measure more, so we can do more" (Apple 2019, 8). Apple explains,

> We take responsibility for our entire carbon footprint. That includes the emissions beyond our direct control, like those from sourcing materials, making our products and our customers using their devices. We calculate our carbon footprint in five major areas: corporate facilities, product manufacturing, product use, product transportation, and product end-of-life processing. We use this information to tell us where to focus.

> **(Apple 2019, 8)**

Businesses should record information to track performance, conformance to objectives and targets, and relevant operational controls. Monitoring equipment must be calibrated and maintained, with records retained for a specified period of time. There also needs to be a periodic evaluation of compliance with relevant environmental legislation and regulations.

Self-monitoring is an important concept that was studied carefully by the NCPL Commission. Specially trained compliance personnel can employ self-monitoring techniques for high-risk operations that are either subject to significant regulations or where the nature and history of such operations or facilities has suggested a significant potential for an industrial accident (NCPL 1996). Self-monitoring may also serve as a training mechanism, as well as a check and deterrent in those situations where reportable performance measures bear on compliance levels. Self-monitoring can be used as an early warning system to alert management of certain types of environmental problems that require changes in procedures, operations, or processes.

For small companies, self-monitoring may simply mean regular attention by line managers to environmental compliance issues on a day-to-day basis, which includes managerial oversight of subordinates (NCPL 1996). Larger companies can afford more sophisticated systems, which include monitoring mechanisms built into operational controls. Companies may wish to consider real-time monitoring for particularly important environmental aspects of operation-like emissions or toxic waste handling procedures. Real-time monitoring may mitigate harmful effects of past noncompliance and prevent repetition of the same mistakes (NCPL 1996). Whatever the level of financial commitment, self-monitoring should be a nonintrusive method of managing risks and reducing potential liabilities.

9.7.29 Nonconformance and Corrective and Preventive Action

Environmental management systems should require businesses to define responsibility and authority to respond to and investigate nonconformance and to take action to mitigate the impacts of nonconformance. The business must initiate and complete corrective and preventive actions to eliminate the causes of actual or potential nonconformance. We call this the identification and elimination of liability-creating conduct. Management systems response actions must be commensurate with the environmental impacts. These actions should result in the implementation and recording of procedural changes in the environmental management system resulting from the actions.

Preventive actions need to be taken in the first instance to avoid noncompliance. Participation in the EPA's voluntary programs, civil and criminal litigation awareness programs, and self-monitoring can remove many of the causes of noncompliance in the course of the development of a sound environmental management system. If an incident occurs despite all these efforts, steps must be taken immediately to minimize risks to the company.

The NCPL Commission found that companies must be proactive in this approach to dealing with incidents of noncompliance before the incident occurs. For example, a company should designate a specific individual with the responsibilities of identifying and responding to actual or suspected violations. That individual should be designated in advance of occasions for such investigations and have special procedures for gathering evidence of misconduct, as well as adequate resources available to conduct an investigation. The company should ensure the independence of the investigators from line managers whose activities or supervision may be subject to investigation. Investigators need to take steps to assure the accuracy and reliability of information gathered during the investigation and conduct interviews in a manner that is likely to preserve the attorney-client and work product privileges.

If evidence of an environmental violation is found, the company needs to make a determination whether to disclose the potential violation to authorities. The EPA and many state regulatory authorities have strict procedures including time limitations for making such disclosures to authorities. Hence, it is important to designate decision-making responsibility and authority for determining when and how decisions to self-report detected misconduct should be made, giving sufficient latitude to the decision makers to take each incident on a case-by-case basis to avoid circumstances where a company's procedures require an action that is not warranted by the facts. Persons responsible for deciding whether to disclose have only a brief period of time, perhaps as few as ten days, to make the decision. During this time, evidence has to be collected, people interviewed, and legal decisions made regarding whether a violation has actually occurred and whether the advantages of reporting the misconduct outweigh the risks of keeping that information confidential.

In analyzing the difficult and delicate factual and legal considerations with a wide range of possible negative consequences, it is also necessary to determine the appropriate scope of disclosure if a decision is made to disclose; address potential conflicts of interest between the business and its employees, who may be subject to charges and who are outside the protection of government disclosure policies; and determine how to preserve legal privileges in the course of making the disclosure to authorities. It may also be necessary to decide whether to cooperate with external investigators and to determine when and how to remediate the environmental harm. Other legal considerations include identifying the scope and ramifications of the business's vicarious responsibility to third parties for the detected misconduct. With all of these decisions occurring within a brief period of time, it is important for the business to lay out the procedures in advance in order to minimize confusion, uncertainty, and inconsistent decisions if an incident arises that requires a decision whether to disclose an environmental violation to authorities.

Reliable records of the investigation and the remediation decision must be preserved to document the nonconformance, as well as the corrective and preventive action taken. In this regard, findings, conclusions, and recommendations reached as a result of measuring, monitoring, auditing, and other reviews of the environmental management system should be documented and the necessary corrective and preventive actions identified. Documenting the noncompliance is included so that evidence may be preserved for future use.

9.7.30 Records

Procedures need to be documented to identify, maintain, and dispose of environmental records. This requirement needs to be carefully distinguished from environmental management system documentation, which involves maintaining in paper or electronic form the records that document the existence of that system. "Records" applies to all environmental records. It provides that records must be legible, identifiable, and traceable to the activity, product, or service of the business that involves environmental aspects and impacts. The records must be readily retrievable; protected from damage, deterioration, and loss; retained as specified; and demonstrate conformance to the standard.

Generally, environmental records will include the company's permits, training records, monitoring data, material safety data sheets and product information, information pertaining to suppliers and contractors, environmental audits, and management reviews. Records of environmental aspects and associated impacts and legislative and regulatory requirements should also be included, together with inspection, calibration, and maintenance activities, as well as details of nonconformance, including incidents, complaints, and follow-up action.

Records management requires developing a means of identification, collection, indexing, filing, storage, maintenance, retrieval, retention, and disposition of all pertinent environmental records, including environmental management system documentation. Careful attention has to be paid to maintaining the integrity and confidentiality of these records in this process.

9.7.31 Environmental Management System Audits

An environmental management system needs to be audited on a periodic basis to ensure that it is being implemented effectively. Businesses need to develop procedures for audit scope, frequency, methodologies, responsibilities, and requirements. This program must include periodic audits to determine if the environmental management system conforms to planned arrangements. The audit must determine if the environmental management system is properly implemented and maintained. The audit program, including the schedule, must be based on environmental impacts and past results. Results of the environmental management system audit must be provided to management. A continual improvement process should be applied to an environmental management system to achieve overall improvement in environmental performance.

The NCPL Commission found that auditing is one of the reasonable steps for a business to take to achieve compliance with its self-imposed standards as well as requirements of law. Evaluative auditing programs can determine the effectiveness of the environmental management system. Audits can be designed to perform different purposes, such as an unannounced audit to deter and detect willful misconduct or preannounced audits, which are likely to be less disruptive and detect unintentional mistakes. Auditing can be done in conjunction with self-monitoring and regular reporting functions of the company. Auditing may uncover a violation that may result in the company's need to decide how to correct the problem and whether to disclose the violation to the government to avoid penalties. The NCPL Commission provides as an example that an environmental audit can check a facility's emissions and the integrity of the monitoring processes in place.

Companies need to evaluate the desirable frequency and scope of audits and assess the independence and reliability of those individuals who perform audits. An open question is whether the company should have audits conducted by persons inside or outside the business or a particular business unit. The NCPL Commission advised that systems need to be devised to assure follow-ups to register findings which have as a component a means for employees or agents to report violations of the standards. For those employees who do report violations, a system should be in place for them to do so anonymously, without retaliation, if they so desire. The audit program's procedures must be fully documented and include a records retention component.

The following companies have publicly disclosed that they consider auditing an important part of their efforts to manage environmental risks.

Apple

Apple sets forth in Appendix C of its 2019 Environmental Responsibility report their Assurance and Review statements. This includes site visits to Apple facilities and the corporate offices; interviews with relevant personnel; a review of the internal and external compliance documentation; an audit of environmental performance data; and a review of "information systems for collections, aggregation, analysis and internal verification and review of environmental data" (Apple 2019, 69).

HP

HP's 2018 *Sustainable Impact Report* has approximately 100 pages of disclosures (47–144), including Audit Results. In the 20-plus years we have been analyzing the results of environmental audits, companies like Apple and HP have gone far beyond what is required to conduct a sound environmental management system.

Nissan

Nissan's Sustainability Report contains a discussion of its "audit system" under which "the Board of Statutory Auditors oversees the Board of Directors. The Statutory Auditors attend board meetings and other key meetings and carry out interview with board members to audit their activities." They supplement the activities of the Independent Auditors who report the third-party assurances audit (Nissan 2018, 114).

IBM

IBM conducts rigorous audits and self-assessments of its compliance with its environmental policy, measures progress of its environmental affairs performance, and reports periodically to the Board of Directors.

Westinghouse

Westinghouse began a formal auditing program in 1988 to perform internal assessments of all facilities and provide management with verification that they are in compliance with government regulations and company policies and procedures. Environmental, safety, and industrial hygiene issues are reviewed to ensure that deficiencies are identified and promptly corrected. The Westinghouse Environmental Audit Program has developed a "world-class" reputation for quality and effectiveness.

Xcel

Xcel describes in detail how its Operations, Nuclear, Environmental and Safety Committee proceeds with its factual reviews; how executive compensation is tied to achievement of carbon emissions reduction goals and financial performance; and how the EMS operates with communications audits and internal and external operations audits (Xcel 2019, 12, 96-97).

Waste Management

This company describes a comprehensive EMS with detailed compliance audits, root cause failure analysis, audits of objectives and targets and regulating violations (WM 2018, 168-70).

9.7.32 Management Review

Top management must review the environmental management system to ensure continued suitability, adequacy, and effectiveness. The necessary information must be collected to allow for a meaningful review. The review must be documented; must consider audit results, changing circumstances, and commitment to continual improvement; and must address the need for changes to policy, objectives, and environmental management system elements.

The review of the environmental management system should include a review of environmental objectives, targets, and environmental performance; findings of the environmental management system, self-monitoring, and audits; an evaluation of its effectiveness; an evaluation of the suitability of the environmental policy and the need for changes in the light of legislation, expectations, requirements of interested parties, the products or activities of the business, or advances in science and technology; lessons learned from environmental incidents; market preferences; and reporting and communication. The following companies have disclosed how they conduct management reviews:

- **Chevron** has agreed to expand "corporate environmental compliance reviews to include assessing the effectiveness of environmental, safety, fire, and health management systems, in addition to maintaining legal compliance. This should help the Company identify and modify those processes, products and practices that may involve unacceptably high risks when judged by the standards of anticipated laws and regulations."

- **StorageTek's** "executive management will regularly review safety and environmental management performance and compliance and ensure adherence to established goals and policies."
- **Lockheed Martin** "reviews company program effectiveness in implementing corporate policy, achieving and maintaining compliance with laws and regulations, stewardship of assets and reduction of EHS costs."

The NCPL Commission found that management reviews, such as those of Chevron, StorageTek, or Lockheed Martin, may wind up disciplining or possibly retraining responsible employees and identifying root causes of noncompliance, including weaknesses in detection practices. External evaluators can provide management with objective evaluations of the environmental management system and provide useful insights on how the program can be strengthened. Management reviews should be conducted regularly and promptly to assure effective follow-up measures.

Many businesses that institute ISO 14001–compliant environmental management systems desire to gain the international recognition earned by successful implementation of their environmental management system. After a business has developed and implemented its ISO 14001 environmental management system, it can retain a third-party registrar to assess the system's conformance with the standard.

References

3M. *2019 Sustainability Report*, 2019. https://www.3m.com/3M/en_US/sustainability-us/annual-report/.

Apple. *Environmental Responsibility Report*, 2018. https://www.apple.com/ca/environment/pdf/Apple_Environmental_Responsibility_Report_2018.pdf.

Apple. *Environmental Responsibility Report*, 2019. https://www.apple.com/environment/pdf/Apple_Environmental_Responsibility_Report_2019.pdf.

Baxter. *2018 Corporate Responsibility Report: Making a Meaningful Difference*, 2018. https://www.baxter.com/our-story/corporate-responsibility.

Bayer. *Annual Report 2018*, 2018. https://www.annualreport2018.bayer.com/.

Bristol-Myers Squibb. *2018 Global Citizenship Report*, 2018.

British Petroleum (BP). *Sustainability Report 2018: Responding to the Dual Challenge*, 2018. https://www.bp.com/en/global/corporate/sustainability.html.

Carver, T.B., and A.A. Vondra. "Alternative Dispute Resolution: Why it Doesn't Work and Why It Does." *Harvard Business Review*, (May/June, 1994).

CH2M. *2017 Sustainability and Corporate Citizenship Report*, 2017. http://www.jacobs.com/sites/default/files/content/basic_page/attachments/ch2m-2017-sustainability-and-corporate-citizenship-report.pdf.

Chevron. *Corporate Responsibility Report*, 2018. https://www.chevron.com/corporate-responsibility/reporting.

ConocoPhillips. *2018 Sustainability Report*, 2018. http://www.conocophillips.com/company-reports-resources/sustainability-reporting/.

Dauer, E.A. *Manual of Dispute Resolution*. New York: McGraw-Hill, 1994.

Dean, M. "Contaminated Sites: Connecticut moves Toward Private ADR." *Dispute Resolution Journal* (January/February 1996).

Dow. *2018 Dow Sustainability Report*, 2018. https://corporate.dow.com/en-us/science-and-sustainability/reporting.html.

DuPont. *Global Reporting Initiative Report*, 2018. https://www.dupont.com/content/dam/assets/corporate-functions/our-approach/sustainability/commitments/product-stewardship-regulator/articles/documents/DuPont%20GRI_Report_2018i.pdf.

EPA. "Superfund Enforcement Mediation, Regional Pilot Project Results, 22E-2201." (October 1991).

EPA. *Enforcement and Compliance Assurance Accomplishments Report FY 1995*, July, 1996.

EPA. *Fiscal Year 2016 EPA Enforcement and Compliance Annual Results*, December 19, 2016. https://archive.epa.gov/epa/sites/production/files/2016-12/documents/fy16-enforcement-annual-results-data-graphs.pdf.

Evonik. *Sustainability Report 2017*, 2017. https://corporate.evonik.com/en/media/publications/cr-report/.

Exelon Corporation. *Environment Policy*. 2017. https://www.comed.com/SiteCollectionDocuments/SafetyCommunity/corporate-environmental-policy.pdf.

Gaynor, K.A., P.D. Kamenar, E.D. Muchnicki, and P. Thomson. "Doing Time FOR Environmental Crimes." *Environmental Forum*, (June 23, 1993).

General Motors (GM). *2018 Sustainability Report: Transformation in Progress*, 2018. https://www.gmsustainability.com/.

Graham Partners. *Sustainability Update*, 2017.

Graham Partners. *2018 Sustainability Report*, April 2019. https://www.grahampartners.net/media-center/.

Hewlett-Packard (HP). *Sustainable Impact Report 2018*, 2018. https://www8.hp.com/h20195/v2/GetPDF.aspx/c06293935.pdf.

Hitachi. *Hitachi Sustainability Report 2018*, 2018. http://www.hitachi.com/sustainability/download/pdf/en_sustainability2018.pdf.

Hongkong and Shanghai Hotels, Ltd. *Corporate Responsibility and Sustainability Report*, 2017. https://www.hshgroup.com/en/sustainable-luxury/sustainability-reports.

IBM. 2018. *2018 Corporate Responsibility Report*. https://www.ibm.org/responsibility/2018.

Inter-Continental Hotels (IHG). *Responsible Business Report 2018*, 2018. https://www.ihgplc.com/responsible-business.

Jacobs. *2018–2020 Sustainability Strategy*, 2019. http://www.jacobs.com/sites/default/files/2019-04/Jacobs-2018-2020-Sustainability-Strategy.pdf.

Johnson & Johnson. *2018 Health for Humanity Report*, 2018. https://healthforhumanityreport.jnj.com/.

Levin, M.E. *Prosecution of Environmental Crimes in Massachusetts*. Massachusetts Department of Attorney General, 1991.

Lockheed Martin. *2018 Sustainability Report: The Science of Citizenship*, 2018. https://sustainability.lockheedmartin.com/sustainability/index.html.

Mackey, J. *Conscious Capitalism*. Cambridge, MA: Harvard Business Review Press, 2013.

Mcinerney, F., and S. White. *The Total Quality Corporation*. New York: Truman Talley Books/Plume, 1995.

Molson Coors. 2018. *2018 Sustainability Reporting: Beer Print Report and Environmental, Social and Governance (ESG) Report*. https://www.molsoncoors.com/en/sust ainability/overview/sustainability-reporting.

Muchnicki, E.D. "Only Criminal Sanctions Can Ensure Public Safety." *Environmental Forum*, June 23 1993.

National Center for Preventive Law (NCPL). *National Center for Preventive Law Corporate Compliance Principles*, 1996. http://www.preventivelawyer.org/con tent/corp_compliance.htm.

Nissan. *Sustainability Report 2018*, 2018. https://www.nissan-global.com/EN/SU STAINABILITY/LIBRARY/SR/2018/.

Schiffer, L.J., and R.L. Juni. "Alternative Dispute Resolution in the U.S. Department of Justice." *Natural Resources and Environment* (Summer 1996).

Schotland, S.D. "Mediation and Arbitration of Toxic Product Liability Cases." *BNA Analysis and Perspective*, August 9, 1995.

Thornburgh, R. "Our Blue Planet: A Law Enforcement Challenge." *Keynote Address to the 1991 Environmental Law Enforcement Conference*, New Orleans, LA, January 8, 1991. https://www.justice.gov/sites/default/files/ag/legacy/2011/08/23/01-08 -91.pdf.

U.S. Department of Justice. "Bureau of Justice Statistics." 1984.

Waste Management (WM). *2018 Sustainability Report: Driving Change*, 2018. https:// sustainability.wm.com/.

Westinghouse. *EHS and Sustainability Policy*, 2017.

Westinghouse. *EHS and Sustainability Policy*, 2018.

Xcel Energy. *Destination 2050: Building the Future*, 2019. https://www.xcelenergy.com /company/corporate_responsibility_report.

Xerox. *2018 Corporate Social Responsibility Report*, 2018. https://www.xerox.com/cor porate-social-responsibility/2018/index.html.

Xerox. 2019. *Xerox Environment, Health, Safety, and Sustainability Policy*. https://www .xerox.com/en-us/about/environment/environmental-policy.

10

Practical Methods to Reduce Litigation Costs

John Voorhees

A corporate risk reduction strategy will be ineffective unless it has a mechanism to reduce the threat of environmental litigation. For any small- to mid-sized business, environmental litigation can drain a business's resources. Even large businesses that have gone the distance and tried environmental cases to completion sometimes refer to them as bet-the-company cases. A typical environmental case may last five years or more from the time of filing until disposition in the district court, resulting in lengthy appeals and even lengthier contribution actions against others who might be responsible parties.

The best way to avoid litigation with the government or private parties is to be in compliance 100% of the time, to have no accidents, and create no wastes. This is a nearly impossible task to accomplish for a company of any size, but particularly for large companies that have far-reaching environmental impacts and responsibilities in multiple jurisdictions. Obviously, an effective environmental management system is the optimal means to avoid environmental disputes and thereby minimize environmental litigation risks. Before any action is taken to control risk, companies must first understand and quantify the risk present, which tends to be an arduous task that too often gets pushed to the back burner.

In 1736, Benjamin Franklin asserted that "an ounce of prevention is worth a pound of cure." Over 20 years ago, the Corporate Counsel Section of the New York State Bar Association issued a report, together with a litigation management model, aimed at showing businesses how to avoid litigation and reduce its attendant costs whenever it occurs (Haig 1997). The report emphasized the need to consult with counsel before a dispute arises to prevent litigation or, alternatively, to simplify protracted legal battles. Preventive law has gained prevalence in the recent past, and just as doctors are increasingly recognizing the importance of prophylaxis, many law schools, lawyers, and judges are beginning to understand how certain preventive measures can be taken to avoid legal disputes.

10.1 Prevention Measures

The concept of prevention needs to be written into environmental risk management and risk reduction systems. The multiple methods of preventing problems need to be set forth with an alternative dispute resolution (ADR) component in the system if prevention fails and a dispute arises. Although ADR is not the perfect replacement for preventive law, it has come a long way over the past 30 years and has the potential to be a valuable supplement to the current legal system. The Department of Justice and the Environmental Protection Agency (EPA) have, on many occasions, supported the use of ADR to minimize and avoid disputes. Before such disputes arise where the government is a party, managers and employees need to be taught, in the first instance, the many procedures that can be used to facilitate problem identification and resolution. These can include interactive training models to teach employees how to consider collaborative decision-making techniques in solving everyday problems before disputes arise, as well as using ADR techniques such as mediation or arbitration when problems mature into disputes but before parties are polarized and headed for the courthouse. ADR techniques are discussed in more detail in other sections of this book. There are other ways to eliminate litigation risk by teaching employees how to avoid disputes in the first instance. Involving counsel and other expert consultants and communication specialists in training and role-playing exercises can improve corporate awareness of the range of possibilities that can be experienced by the business before disputes occur.

In particular, legal counsel can further assist companies with the development of litigation awareness programs to identify and eliminate liability-causing conduct. Specifically, targeted risk reduction mechanisms can be discussed, materials can be disseminated, and tests or exercises can be created to motivate litigation cause awareness. Seminars and other types of interactive training can be conducted for management and employees on a variety of topics, including employment law, environmental compliance, cyber security, and unfair trade practices—to name a few of the most litigated areas of the law. These techniques focus on the identification and elimination of liability creating conduct.

A litigation awareness program can demonstrate how to implement a legal and regulatory compliance program designed to reduce the risk of litigation being filed by an employee whistleblower or a third party against the business. Educational and interactive simulations, case studies, and programs can also teach employees how to avoid creating emails and written records or oral reports that can be taken out of context and used against the business. This would not be an obstruction of justice course, but rather a technical and legal session on teaching employees not only how to avoid breaking the law but also how to avoid creating documents, especially emails and casual exchanges of information, that make it look like they broke the law. When documents are created, document control procedures need to be carefully implemented so that the business can maintain its compliance record in an organized and retrievable manner and prevent the destruction of relevant documents that could be used affirmatively to prevent a claim from being filed. Information control procedures should be created for this and other purposes. Counsel can also review electronic information and records, including agreements, to assure that they include litigation avoidance mechanisms, such as arbitration and mediation clauses, to prevent unnecessary litigation over commercial and other types of disputes that frequently arise despite the best intentions of the business and its employees. And, most importantly, systems can be developed to reduce cyber security risks and exposures for businesses that rely heavily on third-party vendors (Grillo 2019).

10.2 Collaborative Decision Making

Many environmental lawsuits have been brought by public interest organizations and by assemblages of citizens. These cases have been filed either against the government to block federal, state, and local permitting decisions or to ensure that representative government meets statutory or regulatory responsibilities, time tables, and objectives. In addition, environmental lawsuits are filed against companies to prevent or be compensated for environmental contamination resulting in natural resource damages, wildfires, floods, and other disruptions of the ecosystem. Litigation avoidance in this context involves the ability of government officials or business leaders to meet with environmental groups and citizen leaders to engage in collaborative decision making to resolve disagreements before they ripen into legal disputes. Local, regional, and national services are available to assist companies, communities, government agencies, environmental groups, and others to resolve problems before a claim arises.

Scientists, facilitators, mediators, public participation and communications consultants, risk assessors, public policy specialists, environmental engineers, and lawyers can provide a team approach to avoid environmental disputes or to resolve them before they mature into litigation. Collaborative decision making provides an open forum for all stakeholders to be involved proactively in a decision-making process before positions are polarized or in anticipation of conflict. History has taught us that there are classic situations where dispute resolution is very difficult, and more current situations such as the Keystone Pipeline (an overview of which is available at https://ballotpedia.org/Keystone_XL_Pipeline) continue to serve as exemplars of the challenging nature of attempted dispute resolutions. The Keystone Pipeline project has been heavily debated among political, social, and environmental groups for the better part of the beginning of the 21st century. As politics shift over time, so do the suggested resolutions regarding the Pipeline. In November 2015, then-President Barack Obama denied TransCanada the permit necessary to begin construction. Just sixteen months later, President Donald Trump signed a presidential permit to approve the pipeline. The proposed pipeline has resulted in hundreds of protests, firmly entrenched positions both for and against the pipeline, and significant litigation well before pipeline construction began. Such a notable example of dispute resolution proves just how heated such topics can become. However, other less intractable problems can be solved with collaborative decision making involving stakeholders who participate in defining the problems and in finding mutually satisfying solutions. The process also includes steps of conflict analysis, process planning, management, and facilitation. If this process fails or a dispute has already been filed in court, ADR techniques can be used to avoid protracted environmental litigation.

10.3 Financial Implications of Litigation

When a dispute erupts that cannot be resolved without resorting to litigation, businesses should fully weigh the financial and human resource implications of long-term environmental litigation. Twenty years ago, when the first edition of this book was published, we highlighted the extraordinary defense costs in the first three years of pretrial litigation at Rocky Flats. We noted Dow Chemical and Rockwell International spent over $20 million in discovery-related expenses for a property diminution case involving a DOE weapons facility in Golden, Colorado. Costs in three other cases involving DOE weapons facilities total nearly $50 million. *A Civil Action*, by Jonathan Harr (1995), and *Toms River, a Story of Science and Solution*, by Dan Fagin (2013), detail the chilling account of citizens who brought toxic tort cases against major corporations, resulting in huge costs and unsatisfactory results for all involved.

Has anything changed? Indeed, experts and lawyers have gotten more expensive. Frequently, the government has become more litigious both in defending its Second World War and Cold War environmental liability for the environmental damages caused by its defense program, and in engaging in "battles of the experts" in contested litigation. Native American tribes have proven to be formidable adversaries in the West, challenging incidences of environmental degradation on reservations such as the Gold King mine disaster, allegedly caused by EPA and its contractor, and pursuing other native rights lawsuits involving reservation land.

Facing litigation damages awards in the hundreds of millions of dollars in the 21st century, large companies need to incorporate into their environmental management systems mechanisms to reduce the risk of protracted litigation whenever possible. This is in addition to mechanisms we have discussed to identify and eliminate liability-creating conduct in the first instance. When litigation begins, companies need to monitor litigation costs closely and review options and strategies continuously as the litigation progresses to trial. A settlement strategy must be developed as early as possible and should be reviewed periodically as more facts become known in the pretrial discovery process.

Trials are getting longer, more frustrating, and certainly more expensive. In the first edition of this book, we featured the Occidental Chemical case involving Love Canal as a classic example. Love Canal came to national attention in 1978 when chemical compounds began leaking into basements of homes abutting the site. The government relocated over 1,000 families. Occidental's predecessor, Hooker Chemical, was found strictly liable and the government pursued punitive damages against Occidental. After 10 years of pretrial proceedings, the trial began in October 1990; 166 witnesses had already been deposed. The liability trial on the punitive damages theory of the government lasted eight months. There were approximately 4,000 exhibits and 81 witnesses (50 lay and 31 expert). Final arguments took place on February 12, 1992, and the judge ruled for Occidental over two years later.

There have been many cases since Love Canal. Settlements reduce expenses, but environmental costs are hideously expensive. Deep Water Horizon and Volkswagen are two extreme examples. Defense fees and costs in the Volkswagen emissions case exceeded $300 million (Reuters 2017). While accidents continue to happen even to well-prepared companies, and notable good corporate citizen companies fall victim to internal mischievous and illegal profiteering, smart companies that are environmentally savvy and socially aware get way out in front of this problem.

10.4 Ways to Minimize Litigation Expenses

While the facts of every environmental dispute will be different, the basic approach to minimizing litigation expenses will be similar whether defending a government enforcement or cost recovery action, a citizen's suit, or a private-party action. To achieve realistic goals, a comprehensive litigation management plan should first be prepared. In-house counsel, or management if the business does not have internal counsel, needs to identify the company's overall goals and objectives for the litigation and create a preliminary cost-benefit analysis. These tasks should be accomplished before or contemporaneously with hiring outside counsel, if necessary, to handle the litigation. The plan needs to be approved and endorsed by management and handled carefully to protect the confidentiality of the strategy. The litigation management plan should include a discussion regarding concerns about adverse publicity and the negative impacts of litigation as well as the benefits to be derived from the company having its day in court.

After the plan is developed, the business can use its retained law firm or in-house counsel to handle the case, or it can consider soliciting requests for proposals and presentations from several law firms and then select the best-qualified counsel to provide the desired legal services (Haig 1997). A company can seek from applicants alternatives for handling the litigation as well as a description of estimated costs for each strategy to meet the company's overall litigation objectives. Alternative fee arrangements are popular with both law firms and clients seeking to share the risks of litigation where outcomes are not assured. Assessments of legal expertise and specialized practice need to be carefully evaluated. All of this can be set forth in a privileged and confidential letter or memorandum that can further clarify the company's expectations from outside counsel.

Once counsel is assigned to the case, the litigation management plan can be expanded to incorporate his or her strategy as to how the case will be staffed and handled. Depending upon the type of case filed, this strategy can be enhanced by taking into consideration the burden of proof on each element of the claim, the likelihood of injunctive relief being awarded, a liability assessment, imposition of fines or penalties, and in Comprehensive Environmental Response Compensation and Liability Act (CERCLA) cases, for example, a determination regarding recovery of response costs and allocation of those costs. The best way to reduce litigation expenses and meet expectations in a cost-effective manner is to prepare a comprehensive, privileged, and confidential litigation management plan and revise it periodically as circumstances demand. Settlement options should be discussed and weighed as the litigation proceeds toward trial. The following illustrates how to create an effective litigation management plan in the context of a cost recovery or contribution action filed under the CERCLA statute.

10.5 Reducing Litigation Costs under CERCLA

When the government brings a CERCLA cost recovery action or a private party brings a contribution action, courts generally bifurcate the case into a liability phase and a damages phase. In the liability phase, the burden is always on the plaintiff to prove a release or threatened release of a hazardous substance by a responsible party. The plaintiff must also prove that he or she has incurred costs. In a government cost recovery action, the burden then shifts to the defense to prove that the government's response costs have been incurred inconsistently with the National Contingency Plan (NCP). In a private-party contribution action, the burden remains with the plaintiff to demonstrate that its costs were incurred consistently with the NCP.

In the prelitigation setting, a business may have already become deeply involved in developing a defense or other risk management strategies. To be able to accurately predict the future course of litigation, not necessarily its outcome, a litigation management plan can be developed to quantify how much it will cost, how long it may take to try the case, and what the possible settlement options are. The plan has to focus initially on the following question: If the business is found liable, what will its share of the costs be? This question may well arise long before the case is filed. For example, the EPA may send a CERCLA *Section 104(e)* letter to the business seeking information about the waste disposal at the site, and the business may subsequently learn what amounts of waste other businesses contributed to the same site.

As a general rule, joint and several liability is often imposed in CERCLA cases, but it is not mandatory. Businesses can reduce liability if they can prove that they only contributed a certain percentage of hazardous waste at a site. An analysis will determine if it is possible to (1) reduce the total amount of liability to an allocable share and (2) quantify how much liability exists. In this process, it will be important to determine the business's percentage of contribution of waste material to the site, the business's rank in relation to the possible contribution of other PRPs, and the relative toxicity of the waste contributed in relation to other PRP wastes. An allocation analysis is a critical factor in determining how much a business actually is responsible for environmental problems. Allocation specialists are available to conduct these analyses. The answer to liability and allocation questions may result in a decision to settle the case if the business is a small contributor, or even a major player. Thus, in determining how to effectively manage a CERCLA case, the first important question to answer is whether the business should litigate at all.

Next, the plaintiff's (government's) best case needs to be addressed on an element-by-element basis. An analysis should be prepared of the probability that a court following a bench trial will conclude that the business is liable to compensate the plaintiff or government for past and future response costs. A detailed summary of the plaintiff's expected proof needs to be prepared to determine the relative strength of its case. In determining how hard the plaintiff will press their case against this particular business, it must factor in what other businesses may be liable for and if they are able to pay an allocable share of the response costs.

For years courts have narrowly construed CERCLA's liability defenses (act of God, act of war, or act or omission of a third party) and have provided very few and limited opportunities for businesses to escape liability or reduce liability. Some defenses have focused entirely on whether a "disposal" or a "release" occurred within the meaning of the statute or whether the site was a "facility." Each statutory defense must be reviewed to determine the likelihood that the court will accept it.

Before commencing the defense, in-house counsel and/or senior management must consider:

- What will be the opportunity, if any, to develop the defense during discovery? (The government has statutory authority to limit discovery.)
- How much will it cost in lawyers' and experts' fees to prepare and present the defense?
- Will there even be an opportunity to present the defense in a liability trial?

Following this analysis, if the business's defense is not compelling, settlement options should again be reviewed and possibly pursued.

For almost 40 years in multi-party CERCLA actions, co-defendants have organized at times into a single cooperative force. Co-counsel from several firms have divided research and briefing tasks to ease the financial burden. PRP agreements have been drafted to create a united front. Defendants have developed methods of alternative dispute resolution for internal allocation issues (Helmstetter 1996). Co-defendants have shared the cost of experts to improve trial strategy by using their combined expertise in handling complex environmental litigation.

One possible technique to simplify trial issues is to stipulate liability if there are no valid defenses. This eliminates issues from the case that can cast the defendants in a bad light (Helmstetter 1996). If the decision is made to contest liability and the court concludes that the defendants are liable, then the defendants must prove that the government's costs are inconsistent with the NCP. Far more frequently, courts defer to the EPA's expertise in choosing a suitable remedy and incurring costs in a manner consistent with the NCP. By contrast, private-party plaintiffs in contribution actions have to prove their costs were incurred consistently with the NCP. Early in the history of CERCLA enforcement, there were cases where plaintiffs were unable to obtain contribution because they did not follow the NCP in conducting "CERCLA-quality" cleanups. That has changed with the gradual increase in awareness of the importance of conducting cleanups consistent with the NCP.

Valid reasons sometimes exist for challenging the extent of the government's remedial efforts in a cost recovery case. Experts can be retained to challenge the government's theories of liability allocation and review government documentation regarding the selection and scope of remedial actions. It is not sufficient, however, to merely file expert reports and affidavits challenging individual site assessment and cleanup costs as excessive and unreasonable. Defendants must have clear evidence that the government overreached in failing to consider cost or in selecting a remedial alternative that is arbitrary and capricious. Therefore, a candid assessment must be made as early as possible in the liability phase as to whether there is any chance of prevailing on these cost recovery issues. Most of the time, the best that can be done in the cost recovery phase is to convince the government (a remote possibility) or a court that some of the expenditures are inappropriate and thus not compensable. In some cases where the defendants' experts have made a showing of inconsistency with the NCP, the government has voluntarily withdrawn some response costs.

10.6 Insurance Coverage

As early as possible, an investigation must be undertaken to determine whether insurance policies provide coverage for environmental damage. The historic review should include:

- Comprehensive General Liability (CGL) policies that were purchased before 1970
- CGL insurance policies that were purchased between 1970 and 1985 which contain pollution exclusion clauses
- Post-1985 CGL insurance policies which contain absolute pollution exclusion clauses
- Personal liability insurance coverage (frequently sold with general liability policies and with excess and umbrella policies)
- First-party property damage insurance policies
- Other insurance policies including environmental impairment liability insurance (EIL)
- Insurance policies of predecessor organizations
- Other parties' insurance coverage (Anderson, Devries, and Rodriguez 1994)

When this information is collected, an insurance profile can be prepared by insurance archaeologists as part of the litigation strategy that identifies all insurers, coverage periods, exclusionary clauses, and notice provisions. Litigation risks are reduced if satisfactory insurance coverage can be located to indemnify the business for the costs of defense and damages. Timely and adequate notice to these insurance carriers must be given. Insurance agents are also a valuable source of information and can be used to notify carriers of potential claims. If policies are missing, then a secondary search for evidence of coverage is necessary. This would include invoices, cancelled checks, policy renewals, and correspondence. Insurance archaeologists who specialize in finding copies or evidence of lost policies can be helpful. The existence of comprehensive general liability policies will be significant if it can be proven that they indemnified or partially indemnified the business for environmental liabilities. When the policy profile is created, an assessment to determine whether additional coverage is required to eliminate or transfer environmental liability should be made as part of the risk reduction strategy.

10.7 Locating Other Potentially Responsible Parties

It seems odd that some 40 years after the enactment of CERCLA, it still is necessary for the PRP group, or an individual defendant, to conduct a thorough investigation to locate other PRPs. There are multiple national PRP search firms and consultants who have considerable experience in conducting these sorts of investigations of other companies responsible for past disposal of wastes. Notwithstanding the *Bestfoods* decision, Courts have been broadening the concept of owner/operator liability to allow a business to include in contribution actions other businesses that stand in the chain of corporate succession, as well as the government, in a whole host of defense-related environmental contamination cases. Some courts have narrowly construed some indemnification provisions of contracts that preexisted CERCLA, while others have entirely ignored such provisions. Unless an indemnity provision specifically contemplated environmental liabilities or was so broad that it can be interpreted to have included those liabilities, it will not afford protection from liability. The more parties there are to share liability, the more money the business can conceivably save in the process. An important factor to consider is the cost of conducting a full-scale investigation, which may result in a contribution action against the other PRPs or the government. As always, the likelihood of success on the merits needs to be balanced with the cost of proceeding against other PRPs for contribution.

To avoid the high cost of litigation, individual or group consideration should be given to hiring an allocation consultant or mediator to determine whether a negotiated settlement is possible. Over 90% of environmental cases settle before they go to trial. It is impossible to settle these cases unless counsel has sufficient knowledge of the facts to be able to give the client company competent advice regarding the strengths and weaknesses of the opposing party's case. Unfortunately, the discovery process in environmental cases is slow and cumbersome, and it takes time and money to gain the requisite knowledge to provide useful advice to the client. Any efforts made by the parties to reach consensus regarding the facts and the identification and possible resolution of legal issues will speed up the process of litigation and ultimately save financial and human resources. Businesses need to consider alternatives to litigation during this process in order to adequately address all options in minimizing environmental litigation risk.

References

Anderson, Devries, and Rodriguez. "A Policyholder's Primer on Environmental Insurance Recovery." *Journal of Environmental Law and Practice* 5 (1994).

Grillo, R. "Third-Party Risk Management." *CIO Review,* December 20, 2019. https://networking.cioreview.com/cxoinsight/thirdparty-risk-management-nid-30507-cid-9.html.

Haig, R.L. *Corporate Counsel's Guide.* 2nd edn. Albany, NY: New York State Bar Association, 1997.

Helmstetter, C.H. "Environmental Litigation Against the Federal Government." *Natural Resources and Environment* (Summer 1996).

Reuters. "Volkswagen Ordered to Pay $125 Million in Legal Fees Over Emissions Scandal." *Fortune,* July 21, 2017. https://fortune.com/2017/07/21/Volkswagen-pay-legal-fees-emissions/.

11

How Voluntary Programs and Other Initiatives Can Improve Performance and Lead to Reductions in Environmental Risk

John Voorhees

When this book was first published in the middle of President William Jefferson Clinton's second term, the Environmental Protection Agency (EPA) had identified the compelling need to develop and implement more innovative, effective, and efficient approaches to environmental protection (EPA 1997a). The EPA recognized at that time that command-and-control regulations to achieve environmental improvements are only part of a multifaceted solution to environmental protection (EPA 1997a). Accordingly, EPA developed multiple innovative and effective programs that were designed to encourage the government to work with the regulated community to minimize environmental impacts, reduce risks, avoid environmental liabilities, and perhaps improve performance. As we stated back then, there were a number of advantages for businesses that participated in these types of programs, including recognition, streamlined regulatory permitting, and a reduction in the cost of doing business. Each of the historic EPA programs was designed for partnerships to be developed among the EPA, the states, the businesses, and community stakeholders to find innovative ways to reduce regulatory burdens and positively impact the environment. All of these programs involved or were related to a partial reduction of corporate environmental risk and, to some extent, avoidance of the prospect of environmental litigation. The following is our description of the most well-developed voluntary programs and of what has happened to these programs in subsequent years during the George W. Bush and Barack Obama administrations.

11.1 Environmental Leadership Program

In the mid-1990s, President Clinton's EPA began a one-year pilot program that served as the basis for the creation of the Environmental Leadership Program (ELP). The goals of the program were:

- To better protect the environment and human health by promoting a systematic approach to managing environmental issues and by encouraging environmental enhancement activities, such as biodiversity and energy conservation
- To increase identification and timely resolution of environmental compliance issues by ELP participants
- To multiply the compliance assistance efforts by including industry as mentors
- To foster constructive and open relationships among agencies, the regulated community, and the public

In order to qualify for this program, a business or federal facility had to have a mature environmental management system, as well as a compliance and auditing program. This meant that the environmental management system must have gone through an initial development period. The business must have participated in community outreach and employee involvement programs that fostered relationships among facilities, employees, and local communities. Then, the business had to submit to the EPA facility-wide compliance audit results and environmental management system information obtained within the previous two years. Federal facilities were required to verify that their parent agencies had endorsed the *Code of Environmental Management Principles* (CEMP) and to describe how the applying facility was implementing that program.

When a business or federal facility became a member of the ELP, the program was intended to facilitate an exchange of information. ELP encouraged the implementation of best practices related to environmental management systems and pollution prevention activities. Businesses and federal facilities that successfully participated in the program received:

- **Public recognition:** the EPA would issue certificates of membership in the ELP and would develop programs and activities designed to publicly recognize ELP members.
- **Logo usage:** members could use the EPA-issued logo in facilities and advertising but not on products.
- **Inspection discretion:** the EPA may reduce or modify discretionary inspections.

- **Reduced regulatory burdens:** the EPA may consider in the future expediting members' permits, providing longer permit cycles, and streamlining permit modifications.

The EPA also encouraged businesses to get involved in the ELP program to develop environmental management systems within the framework of ISO 14001. According to the EPA, at the time, the purpose of ISO 14001 was to provide businesses with the elements of an effective environmental management system that could be integrated with other management requirements to help businesses achieve environmental leadership and economic goals. Becoming a member of ELP did not create a conflict with facilities seeking to become certified under ISO, nor did it require certification under ISO. The EPA recognized that ISO 14000 was useful as a management tool that, when integrated into the regulatory system, allowed businesses to obtain comprehensive environmental, economic, and other benefits.

The ELP has been continuously updated and added to. The ELP is a voluntary program that recognizes states and organizations that exceed environmental expectations with regard to sustainability. There are three levels: Bronze, Silver, and Gold. Each state has differing incentives to get businesses to participate. In Colorado, for example, for one year of clean compliance you get one year of Bronze recognition for the environmental project. As of 2015 Colorado had 33 Bronze Achievers, 29 Silver Partners, and 96 Gold Leaders (Colorado Environmental Leader 2015). Since the book was written and the pilot program was launched, states have established their own forms of motivation to conform to regulations and exceed expectations.

11.2 Project XL (1995–2002)

On March 16, 1995, President Clinton created Project XL (eXcellence and Leadership) as part of a 25-point program within his Administration titled *Reinventing Environmental Regulation*. Project XL was originally designed to "give a limited number of regulated entities an opportunity to demonstrate excellence and leadership" (EPA 1997b). The XL project was originally conceived outside of government by a multi-participant process convened by the Aspen Institute (GEMI 1996). The White House adopted the Aspen Institute's findings and placed the XL concept in a White House Policy on *Reinventing Environmental Regulation*. The proponents of the XL program decided to provide regulated industries, governments, and communities with flexibility to develop alternative strategies that would replace or modify specific regulatory requirements on the condition that they produce greater environmental benefits. When the program was initiated, the EPA stated, however, that "[i]n exchange for greater flexibility, regulated entities will be held to a higher degree of accountability for demonstrating project results."

Project XL was controversial from the outset because of its heavy emphasis on accountability and its lack of a uniform approach to assessing environmental benefits and determining how they should be measured. Indeed, the EPA, public interest groups, state and local governments, and business leaders have been trying to reach a consensus on what constitutes superior performance. How this issue is ultimately resolved may well determine how the EPA will reassess and redesign its environmental regulatory program as it modifies its regulatory focus from command-and-control regulations to more flexible regulations designed to provide positive incentives to business to take preventive measures to protect the environment (Orsato 2009).

11.2.1 EPA's Selection Criteria

As of February 4, 1997, only 43 proposals had been submitted to the EPA for review, almost half of which were subsequently withdrawn. Projects were chosen that were able to achieve environmental performance that is superior to what would be achieved through compliance with then-current and reasonably anticipated future regulation. The EPA noted, "Cleaner results can be achieved directly through the environmental performance of the project or through the reinvestment of the cost savings from the project in activities that produce greater environmental results" (EPA 1997b).

The EPA's selection criteria focused on innovative strategies for achieving environmental results. One such proposal was a five-year permit tailored for Merck & Co.'s drug manufacturing plant in Stonewall, Virginia, to replace its coal-fired power plant with one that operated on natural gas, which would result in a 25% reduction in sulfur dioxide and a 10% reduction in nitrogen oxides emissions.

The EPA's preference was for projects that had processes, technologies, or management practices that prevented the generation of pollution rather than controlling pollution once it has been created. The EPA wanted projects that embodied a systematic approach to environmental protection that tested alternatives to several regulatory requirements or affected more than one environmental medium. The EPA preferred projects that were intended to test new approaches that could conceivably be incorporated into the Agency's programs or in other industries or facilities in the same industry.

11.2.2 EPA's Pilot Project Examples

In industry proposals of facility-based XL projects, the EPA suggested that "national environmental requirements may not always be the best solution to environmental problems. Substantial cost savings can sometimes be realized, and environmental quality enhanced, through more flexible approaches involving pollution prevention." Accordingly, the EPA proposed that pilot projects could focus on individual facilities and test alternatives to current environmental management approaches driven by compliance with existing regulations. The EPA sought proposals with overall objectives to devise and test more flexible approaches resulting in both better environmental results and reduced compliance costs (EPA 1997b).

The EPA suggested that some XL projects might focus on national environmental regulations that apply to many industries. It acknowledged that these regulations "are often promulgated piecemeal over a long period of time rather than as a comprehensive environmental program." Thus, one type of project to address this particular problem

> might take the form of combining all federal (and possibly state) require-
> ments for an industry into a single, integrated Final Project Agreement.
> Sector-based and place-based strategies might be combined in a project
> that focuses on a number of facilities in the same or related industries
> within a given geographic region or ecosystem.

The EPA further suggested that "[p]rojects might propose development of enforceable 'best management practices' for pollution prevention or pilot the application of upcoming ISO 14000 voluntary environmental standards within a specific industry sector" (EPA 1997b).

11.2.3 Intel Corporation

Intel's Fab 12 facility, which manufactures semiconductors in Chandler, Arizona, implemented an *Environmental Management Master Plan* that included a facility-wide cap on air emissions to replace individual permit limits for different sources of air emissions. The Final Project Agreement was signed on November 19, 1996. In this agreement, Intel committed to:

- Maintaining air emissions for oxides of nitrogen, sulfur dioxide, carbon monoxide, particulate matter, and volatile organic compounds at a level that ensures the current facility, and any other manufacturing facility built at the site, is a "minor" air emissions source as defined by the Clean Air Act
- Using state health-based guidelines to establish enforceable emissions caps for emissions that affect the community adjacent to the facility; in addition, these health-based standards will be used voluntarily to set emissions levels to increase protection for those working in the facility
- Reducing water consumption and the generation of solid, non-hazardous chemical and hazardous waste
- Establishing property line setbacks 20 times greater than required by local zoning authorities

- Reducing vehicle miles traveled by employees
- Participating in equipment donation and training programs

Intel is the first company to agree to make all its environmental data publicly available on the internet as part of a standard reporting mechanism.

11.2.4 Lean Manufacturing

Lean manufacturing is a systematic method that originated in Japan (stemming from Toyota) in efforts for the industry to minimize waste without sacrificing productivity. It is a business model endorsed by the EPA that emphasizes getting rid of wasteful techniques, creating quality products, and being cost efficient.

> While the focus of lean manufacturing is on driving rapid, continual improvement in cost, quality, service, and delivery, significant environmental benefits typically "ride the coattails" or occur incidentally as a result of these production focused efforts. Lean production techniques often create a culture of continuous improvement, employee empowerment, and waste minimization.
>
> **(EPA 2019b)**

11.2.5 Columbia Paint & Coatings

Washington State Department of Ecology and Washington Manufacturing Services partnered in a lean and environment pilot project that aided the manufacturing facility of Columbia Paint & Coatings. The goals of the pilot project were to develop a "collaborative partnership" between the WA State Department of Ecology and the Washington Manufacturing Services, to evaluate the "benefits and synergies of deliberately integrating environmental tools into on-the-ground lean practices," and to "gain the expertise to offer and promote future lean and environment projects to manufacturers statewide." Through this project's activities, Columbia Paint

> reduced production lead and cycle times, overproduction, material loss and damage, operator travel time, and downtime. The process improvements also reduced raw material wastes, wastewater discharges, volatile organic compound (VOC) emissions, and hazardous wastes. Furthermore, as a result of pilot project activities, Columbia Paint now reuses all wash water from white paints and incorporates it into products. Cost savings for Columbia Paint are expected to total about $210,000 per year.
>
> **(Washington State Department of Ecology 2008, 5)**

11.2.6 General Motors

GM announced that suppliers must create an environmental management system during operations in conformance with ISO 14001. Additionally, GM is "developing a broader supply chain initiative, with involvement from EPA and the National Institute of Standards and Technology (NIST), that some participants hope will become a vehicle to integrate technical assistance on advanced manufacturing techniques and environmental improvement opportunities" (EPA 2019a).

11.2.7 Weyerhaeuser's Flint River Operation

Weyerhaeuser Company's pulp manufacturing facility in Oglethorpe, Georgia, agreed to minimize the environmental impact of its manufacturing processes on the Flint River and surrounding environment by pursuing a long-term vision of a minimum (environmental) impact mill. Weyerhaeuser Company took steps to decrease water use while meeting or exceeding all regulatory targets. The Final Project Agreement was signed on January 17, 1997.

Through a combination of enforceable requirements and voluntary goals, Weyerhaeuser agreed to improve the health of the nearby Flint River and surrounding watersheds by:

- Cutting its bleach plant effluent by 50% over a 10-year period
- Reducing water usage by about 1 million gallons a day
- Cutting its solid waste generation in half over a 10-year period
- Committing to reduce energy use
- Reducing constituents of hazardous waste
- Improving forest management practices in over 300,000 acres of land by stabilizing soil, creating streamside buffers, and safeguarding unique habitats
- Implementing ISO 14001 standards to create an effective environmental management system

11.3 The Common Sense Initiative (1994–8)

On July 20, 1994, EPA Administrator Carol Browner announced the formation of a new program titled the Common Sense Initiative (CSI). According to the EPA, this program was intended to bring together businesses, federal and state governments, environmental and environmental justice groups, and labor to take a fresh look at the way the environment is protected. The key themes of the program were establishing flexible and creative new methods to achieve environmental goals; whole-industry, whole-facility approaches; greater multimedia focus; greater incentives for pollution prevention; stakeholder involvement to identify "cleaner, cheaper, smarter" solutions that are good for both the environment and the economy; and consensus building on how to support the environment and reduce compliance costs.

The EPA had 40 CSI projects underway in the following six industry sectors: automotive manufacturing, computers and electronics, iron and steel, metal finishing, petroleum refining, and printing. According to the EPA's statistics, these six industries comprised over 11% of the U.S. Gross Domestic Product, employed nearly 4 million people, and accounted for 12.4% of the toxic releases reported by all American industry in 1992.

The ongoing projects were:

- A group of metal finishers developing new procedures to reduce regulatory burdens on small finishers in return for superior performance
- A group of petroleum refiners rewriting rules to consolidate, streamline, and simplify air emissions reporting requirements
- Iron and steel manufacturers writing principles on how to clean up abandoned iron and steel brownfields and return them to productive uses
- The computer and electronics industry working on ways to eliminate Resource Conservation and Recovery Act (RCRA) barriers to pollution prevention, recycling, and water conservation
- Auto manufacturers creating more flexible regulatory approaches to reduce burdens on industry, a cleaner environment, and improved community participation in environmental decision making
- The printing industry developing a permit system for printers that provides operational flexibility, reduces pollution across all media (air, water, and land), and improves protection of workers, communities, and the environment

After two years of effort, an oversight group has adopted a series of environmental goals that need to be attained by the initiative. They include:

- Improving information and data collection
- Improving energy efficiency through projects with the U.S. Department of Energy
- Increasing recycling and eliminating barriers posed by RCRA
- Addressing water quality issues
- Improving community-based approaches

- Exploring the green track, involving an alternative regulatory system
- Discovering new methods to implement emerging ideas

In contrast to the ELP and Project XL, the CSI experienced success as a program because businesses were much more accustomed to searching for ways to reduce costs and burdens as opposed to defining in words and actions what is meant by the phrase "environmental leadership." The initiative has clearly helped to incubate some innovative ideas that may ultimately result in "cleaner, cheaper and smarter environmental management strategies" (GEMI 1996). Some environmental regulations have had a beneficial effect on businesses that have been forced to discover new methods to comply with a complex regulatory scheme.

11.4 ClimateWise

ClimateWise combined a number of these theories into a comprehensive program to achieve industrial energy efficiency and pollution prevention. The program was jointly operated by the U.S. Department of Energy and the EPA. It was designed to assist the U.S. in honoring its international commitment to reducing greenhouse gas emissions to 1990 levels by the year 2000. Climate change prevention measures were the focus of international negotiations. In the U.S., 280 ClimateWise participants, representing 6% of industrial energy use, were actively involved in the program. Participants included large businesses, such as DuPont, AT&T, Georgia-Pacific, Fetzer Vineyards, Johnson & Johnson, and General Motors, and smaller businesses like La Junta, Colorado's DeBourgh Manufacturing, which uses environmentally friendly products and technologies to manufacture state-of-the-art lockers for schools and other users throughout the U.S.

The intent of the program was for businesses to develop a partnership with the government to reduce waste and become more energy efficient. According to the director of the ClimateWise program, the government was not seeking to dictate any specific technologies or create any regulatory barriers to reducing operating costs. Rather, through programs such as ClimateWise, the government was providing businesses with free technological advice on altering production processes; switching to lower carbon-content fuels and renewable energy supplies; substituting raw materials; implementing employee mass transit, car pool, or van pool programs; and auditing and tracking energy use for efficient improvements.

ClimateWise began as a voluntary pledge program where businesses seeking to reduce operating costs and gain national recognition agreed to voluntarily reduce pollution. Participating businesses are now encouraged to publish a policy or statement on energy efficiency, establish an energy management team, set improvement targets, monitor and evaluate performance levels, and increase awareness of energy efficiency among employees. Businesses can voluntarily report their emissions reduction through the Energy Policy Act's § 1605(b) database and ensure eligibility for ClimateWise recognition.

11.5 The Merit Partnership for Pollution Prevention

The Merit Partnership for Pollution Prevention (Merit) was a cooperative venture of the public and private sectors to develop and promote pollution prevention practices and technologies that protect the environment and contribute to economic growth (Reich 1997). Merit brought together businesses engaged in manufacturing, banking, insurance, environmental risk assessment, accounting and auditing, and the government to facilitate demonstration projects. These projects showed how environmental management systems can be strengthened and implemented in different industries to achieve both improved environmental performance and economic competitiveness among businesses in the U.S.

Merit was created in 1993 when government and industry leaders were beginning to realize that it was both possible and imperative for government and industry to work together to achieve their respective goals of environmental protection and economic growth (Reich 1997). Merit's first project involved the metal-finishing industry in California, where pollution prevention projects were implemented in seven small- to mid-sized metal-finishing facilities. The companies used innovative technologies to reduce waste generation and recover materials from waste streams for reuse and recycling. The results were published in technology transfer documents, videos, and technical workshops to promote pollution prevention to other metal finishers.

11.6 Green Power Partnership

The Green Power Partnership (GPP) was established in 2001 by the EPA in an effort to protect health and the environment by companies' voluntary use of green power to advance the development and market for renewable electricity sources. Since the inception of the partnership, the voluntary market has grown by nearly 5,000 percent. The program provides participating companies with a framework including benchmarks, market information, technical assistance, and public recognition. In return, the participating partner companies commit to using green power sources for all or a portion of their electricity consumption.

As of January 2018, GPP has more than 1,700 partner companies voluntarily using billions of kilowatt-hours of green power annually. Current partners' green power use represents nearly 40% of the U.S. voluntary green power market. Sixteen percent of all Fortune 500 companies and thirty-five percent of all Fortune 100 Companies are partners with GPP, clearly having seen the benefits the Partnership has to offer (EPA 2018).

11.7 Center for Corporate Climate Leadership

Established in 2012, the Center for Corporate Climate Leadership (CCCL) encourages organizations to identify and achieve cost-effective greenhouse gas emission reductions through innovation. The CCCL serves as a comprehensive resource for companies, assisting them in measuring their greenhouse gas emissions and providing technical tools, guidance, and information-sharing opportunities with other companies working to reduce their climate change impact. The CCCL also co-sponsors the Climate Leadership Awards event, recognizing exemplary corporate and individual leadership in addressing climate change issues. The CCCL is preceded by the EPA's Climate Leaders program, which ended in 2011 (EPA 2016).

11.8 WasteWise

WasteWise is a partnership between businesses, governments, and nonprofit organizations. Partners voluntarily set goals to reduce their waste and incorporate sustainable materials management into their waste-handling process, while simultaneously realizing economic benefits. They then publicly demonstrate these achievements through the WasteWise community and have the opportunity to receive awards and be publicly recognized in WasteWise publications and meetings. National award winners and honorable mentions for 2017 include Kohl's Department Stores, Sears, Frito Lay, and the City of Urbana (EPA 2019c).

11.9 State and Local Programs

In addition to these programs at the federal level, state and local governments have been taking the lead on cooperative programs with businesses. For example, California has implemented the CoolCalifornia Green Businesses program, where businesses receive free advice and technical assistance on meeting their environmental goals (California Air Resources Board). Green Business staff members work with member businesses to ensure that they are in compliance with applicable environmental regulations by conducting site visits on wastewater, storm water, hazardous waste, air quality, and other regulated environmental factors that are of importance to the business. Green Business members also receive their own unique checklists and pledge actions to meet their goals.

Boulder County, Colorado, similarly partners with local businesses to achieve environmental goals and promote compliance. Through their Partners for a Clean Environment (PACE) program, businesses gain access to free expert advisor services, financial incentives, and a certification program to help them measure and gain recognition for their energy, waste, water, and transportation achievements (Boulder County 2019). Currently over 300 businesses and municipal operations in Boulder are PACE members, making up approximately 20% of all businesses in the county. These businesses are able to reduce their costs and improve resource conservation, while also being recognized for their positive impacts on the community and environment.

11.10 Effectiveness of the Voluntary Programs

The most obvious question that arises after reviewing the EPA's voluntary programs is whether any of them resulted in lasting environmental benefits and risk reduction. On March 19, 1997, the EPA's Office of Inspection General (OIG) submitted an audit report to the Agency as to whether two older programs, the Radon and ENERGY STAR® voluntary programs, actually had achieved these two objectives. The OIG concluded that both programs provided an "impetus to overcome the barriers to energy efficiency and change consumer behavior. As a result, they were effective at achieving environmental benefits and reducing health risks" (EPA 1997b). The audit report also provided some insights into how the EPA's other voluntary programs can benefit business and result in similar reductions of environmental risk.

The voluntary Radon Action Program was established by the EPA in 1985 to ensure that the required technical knowledge about this deadly, naturally occurring gas exists and is accessible to homeowners, contractors, and state and local officials. In 1992, the Office of Policy, Planning and Evaluation concluded that while the Radon Program had made some progress in increasing radon awareness and testing, public information alone would not be sufficient to achieve significant, long-term risk reduction. Hence, the EPA recommended radon testing in real estate transactions and building radon-resistant homes as cost-effective approaches in high-risk areas (EPA 1997b).

In 1993, the EPA initiated the ENERGY STAR® program as part of the Climate Change Action Plan to reduce greenhouse gas emissions through voluntary partnerships with businesses and public institutions. The ENERGY STAR® program was designed to get consumers and businesses to use more energy-efficient products. The program had three goals:

- Increase market penetration of existing energy-efficient products
- Ensure that manufacturers' and homeowners' investments in energy efficiency are cost effective and product quality is sustained or improved
- Change consumer behavior

ENERGY STAR® programs focused on encouraging energy-efficient technologies in specific areas such as lighting, office equipment, commercial buildings, and homes.

The OIG audit was limited to the Radon Program and three ENERGY STAR® programs: Office Equipment, Buildings, and Homes. The environmental benefits of the Radon Program showed that in high radon areas, radon awareness was at 78% and testing was at 13%. The ENERGY STAR® program for Office Equipment achieved, by the end of 1995, a savings of 2.3 billion kilowatt-hours of electricity at 1,300 pounds of carbon emission. The OIG stated that

> the role of voluntary programs is to encourage the manufacture and consumers' acceptance of risk reduction in the market phase. By changing consumer behavior and effecting a market transformation, voluntary programs achieve lasting environmental results and reduced risks. As the market is transformed by new environmental innovations, the EPA will be able to reduce its program support and allocate its resources to other products or programs.

(EPA 1997a)

The OIG audit concluded that voluntary, or nonregulatory, programs can be an effective tool for reducing risk and achieving environmental results, as long as they demonstrate several good management practices. Those practices include:

- Planning
- Educating people about incentives
- Providing quality support
- Working with outside organizations
- Evaluating progress and making adjustments

These good management practices enabled the program to achieve environmental benefits of energy savings, pollution reduction, and reduction of radon exposure (EPA 1997a). The audit results will undoubtedly encourage the EPA to continue its efforts to have businesses regulate themselves instead of using old-fashioned command-and-control methods to protect the environment.

Changing human behavior was perhaps the most critically important component of a successful voluntary program. The EPA believed that educating people about incentives is an effective way to get people to act, especially when information about the problem is not enough to get the desired result. "Financial and market incentives are strong motivators for consumers and corporations. The more value corporations and consumers place on the incentives, the higher the rate at which they will take the desired action, thereby decreasing risk" (EPA 1997a).

In 2002, the EPA launched Climate Leaders Program to help members to hit greenhouse gas reductions targets. This voluntary program challenged businesses to reduce emissions through taking inventory of GHG emissions, creating reduction goals, and sending in their progress to the EPA. Additionally, the Climate Leaders work with each other and the EPA to create climate change strategies. As of 2008 only 80 companies of the 153 had announced their goals, unfortunately. Furthermore, only 11 of these companies had achieved the goals. Though this program was not 100% successful, participation in it became a differentiating factor for the businesses and helped to build them credible records as they received EPA recognition as corporate environmental leaders (Orsato 2009, 84).

In order to be successful, voluntary programs need commitments from outside organizations to get businesses and consumers to take a desired action. The EPA has never had and never will have sufficient resources and expertise to effect behavioral changes. It must rely on businesses and other nongovernment organizations that are interested in achieving the same environmental goals. Outside organizations became the key to communicating the Radon Program's message because the EPA's messages on radon had a limited effect on many audiences; the informational materials produced by bureaucracies were often untimely and generic, reducing the number of audiences they reached; the EPA had limited effective channels for sending out the information; and the outside organizations were more connected to the target audiences (EPA 1997a). The OIG's audit validated the EPA's approach of using voluntary environmental programs to create greater awareness of environmental impacts and more opportunities to discover ways to protect the environment.

Since the Clinton-era initiatives and experimentation on voluntary programs to improve environmental performance, the EPA's environmental governance has taken different forms under different federal leadership. George W. Bush's EPA emphasized both corporate self-governance and federal regulation. Bush made cuts to federal funding for pollution monitoring and also shut down some scientific libraries used by the EPA to push for regulations. Bush also pushed for regulations to reduce power plant emissions and address interstate transport of particulates and ozone, although they had limited success in the Supreme Court. Despite this, U.S. emissions of the six most widespread air pollutants were reduced during Bush's tenure. With mixed emphasis came mixed reviews. The general consensus is that U.S. environmental performance improved from 2001 to 2009 regardless of Bush's efforts or lack thereof.

The Obama Administration placed a large emphasis on environmental governance both nationally and internationally. Obama invested over $90 billion in the green economy, including $29 billion for improving energy efficiency and $21 billion toward generation of renewable energy. The EPA experienced many enforcement victories during Obama's time, including the BP Deepwater Horizon oil spill and Volkswagen settlements. Obama's national emphasis on environmental protection influenced the global arena as well. America signed onto the Paris Climate Agreement in 2016, the world's first international commitment to mitigating climate change. America's emphasis on environmental governance as well as the success of environmentally friendly private innovation encouraged other nations to place a similar emphasis on the issue.

The Trump Administration looks much different than that of Obama. Trump pulled the U.S. out of the Paris Agreement just a few months into his presidency. Trump's Executive Order (EO) 13783—Promoting Energy Independence and Economic Growth—revokes Obama's EO 13653—Preparing the United States for the Impacts of Climate Change—as well as federal carbon pollution standards. Projections from the Rhodium Group, an environmental research group, predict that U.S. GHG emissions will cease to decrease and level off sometime in 2019.

References

Boulder County (Colorado). "Partners for a Clean Environment Program (PACE)." 2019. https://www.bouldercounty.org/environment/sustainability/partners -for-a-clean-environment/.

California Air Resources Board. "Be a Green Biz." https://coolcalifornia.arb.ca.gov/ article/be-a-green-biz.

Colorado Environmental Leader. *Colorado Environmental Leadership Program (ELP) 2014–2015*, 2015. https://www.colorado.gov/pacific/sites/default/files/DEHS _ELP_Brochure2014thru2015.pdf.

EPA. "Risk Reduction Through Voluntary Programs." Office of Inspector General (Report of Audit), EIKAF6-05-0080-7100130, March 19, 1997a.

EPA. "Project XL." *EPA 231-F-97-003*, April, 1997b.

EPA. "About the Center for Corporate Climate Leadership." 2016. https://www.epa.gov/climateleadership/about-center-corporate-climate-leadership.

EPA. "Green Power Partnership Program Overview." 2018. https://www.epa.gov/greenpower/green-power-partnership-program-overview.

EPA. "General Motors Corporation." 2019a. https://www.epa.gov/lean/general-motors-corporation.

EPA. "Lean Manufacturing and the Environment." 2019b. https://www.epa.gov/sustainability/lean-manufacturing-and-environment.

EPA. "WasteWise." 2019c. https://www.epa.gov/smm/wastewise#01.

GEMI. *Industry Incentives for Environmental Improvement: Evaluation of U.S. Federal Initiatives*, September, 1996.

Orsato, R.J. *Sustainability Strategies: When Does it Pay to Be Green?* INSEAD Business Press, 2009.

Reich, D. "The Merit Partnership for Pollution Prevention—An Overview." *Paper Presented at the Inter-Pacific Bar Association 7th Annual Meeting and Conference*, Kuala Lumpur, Malaysia, April 30, 1997 (on file with the authors).

Washington State Department of Ecology. *Lean & Environment Case Study: Columbia Paint & Coatings*. 2008. https://www.epa.gov/sites/production/files/2016-11/documents/columbia_paint_coatings.pdf.

12

How Document Control Systems Can Reduce Risk

John Voorhees

12.1 Document Control

There is no better way to shut down a government investigation of an alleged environmental violation than to show a fully documented and verified record of compliance history. A document control system and the management of records are essential components of business operations. Sloppy control procedures can result in much aggravation and expense to the business when documents are not easily retrieved when needed. But daily operational problems pale by comparison to the charge of obstruction of justice when a prosecutor accuses a business of deliberate destruction or concealment of evidence. Document control is an important part of an environmental management system that can result in a significant reduction of environmental risk if set up and handled properly.

Problems are not limited to the rare criminal cases brought against businesses. In one civil case, Texaco discovered that document destruction by rogue employees was a contributing factor to immense civil liability and an onslaught of bad publicity. One employee tape-recorded others making disparaging remarks about minority employees who had sued the company. According to newspaper accounts, the recorded conversations also involved a discussion of destroying possible evidence. The company paid millions of dollars in settlement costs and had its reputation for equal employment severely damaged. A document control system that forms an integral part of a system's operations can address the issues that created Texaco's liability, achieve regulatory compliance, and thus avoid civil and criminal liability. The following is guidance on how to operate an effective document control system.

12.1.1 Environmental Management System Documentation

The original ISO 14001 standard stated under the heading "Environmental Management System Documentation" that businesses shall maintain information in paper or electronic form to describe the core elements of the environmental management system. That standard provided:

> The level of detail of the documentation should be sufficient to describe the core elements of the environmental management system and their interaction and provide direction on where to obtain additional information on the operation of specific parts of the environmental management system.

The documentation can be in the form of a single manual, if desired, and may be integrated with documentation of other systems implemented by the organization. Related documentation can include process information, organizational charts, internal standards and operation procedures, and site emergency plans.

A written manual of the environmental management system can be readily disseminated to employees, suppliers, and other interested parties—and is strong evidence that a company is serious about its compliance efforts. The written policy should address the acceptable behavior of the businesses' employees and state the expectation that every employee will follow the policy.

12.1.2 Document Control

Document control requires businesses to establish and maintain procedures for controlling all environmental documents. A document control system, however, includes procedures that require not just environmental documents but all relevant documents to be preserved for set retention periods until they can be destroyed consistently with laws and regulations and without damaging the interests of the business. Document control requires businesses to establish and maintain procedures for controlling all environmental documents and then how these procedures interrelate to a general records retention system.

The original ISO 14001 standard provided for a system for controlling document records to be established and maintained so that the documents can be located, periodically reviewed, revised as necessary, and approved for adequacy by authorized personnel. Current versions need to be made available at all locations where operations essential to the effective functioning of the environmental management system are performed. Obsolete documents should be assured against unintended use by labeling them as such or by promptly removing them. A decision may be made to suitably identify and retain certain records for legal and/or knowledge-preservation purposes. For all purposes, documents need to be:

- Legible
- Dated (including dates of revision)

- Readily identifiable
- Maintained in an orderly manner
- Retained for a specified period
- Traceable to the activity, product, or service involved

Procedures and responsibilities need to be established and maintained concerning the creation, modification, and disposition of the various types of environmental records. Training records and the results of audits and reviews need to be stored and maintained so that they are readily retrievable and protected against damage, deterioration, or loss. Retention times must be established and recorded.

Document control also requires procedures for identification, maintenance, and disposition of records that are needed for the implementation and operation of the environmental management system, and for recording the extent to which planned objectives and targets have been met. Environmental records can include, but are not limited to, information on applicable environmental laws or other requirements; complaint records; training records; process information; product information; inspection, calibration, and maintenance records; pertinent contractor and supplier information; incident reports; information on emergency preparedness and response; information on significant environmental aspects; audit results; and management reviews.

When an effective environmental document control system is implemented, a business can comply with relevant laws, save substantial sums of money by eliminating valueless papers, and improve access to information for review. A well-crafted document control policy statement assists senior management in understanding the rationale for improving records management procedures. It also ensures that the system, as implemented, achieves broad corporate environmental compliance principles and goals.

A comprehensive environmental document control system cannot be created independently from a business's other document control procedures. Indeed, all sizes of companies already store vast quantities of records for business purposes wholly apart from efforts to achieve environmental management system conformance. The existing records procedures need to be systematically applied to the environmental document control program to ensure that the proper records are preserved for the correct period.

12.2 Practical Steps to Document Control

Here are some practical steps a business can take to create an effective environmental document control system and integrate it into the business's existing records system.

12.2.1 Inventory

A business should begin the process by undertaking a comprehensive records inventory. The inventory should involve all departments to determine what records the business creates on a daily basis in the ordinary course of business, and what records it maintains to ensure that the program covers all major documents.

The records inventory may serve a dual purpose. It should disclose which employees and agents are not exercising caution in their documentation of the business's internal and external business activities. It may also uncover documents most likely to cause concerns regarding potential liability. The inventory can be designed to locate all the records the business generates. It should take into consideration the structure of the organization and the scope of its activities (the states and foreign countries in which it does business), and the agencies that regulate its activities. The types of documents located in the inventory can be summarized in a series by content such as accounts receivables, contracts, invoices, personnel files, quality control, etc. (Skupsky 1994). Finally, all federal and state laws and regulations that apply to the business's activities must be researched to determine what records need to be retained and for how long.

12.2.2 Determination of Retention Periods

After conducting initial research, one expert recommends the identification and use of functional retention categories to establish a legal document retention schedule. He advises that there are five legal categories that affect records retention periods:

- Requirements to keep records, such as certain employment records, that do not state retention periods
- Requirements to keep records for specified retention periods, such as hazardous waste manifests
- Statutes of limitations or limitations of action periods that specify when legal actions or lawsuits can be initiated
- Limitation of assessment periods that specify when taxes may be assessed and tax records audited
- Pending, threatened, or potential litigation and government investigations

(Skupsky 1994)

The business must also determine how long to retain each category of documents.

The next step would then be to assign one of the functional records categories listed above from the records retention schedule to each record listed in the inventory. The records retention period for the functional retention category also becomes the record retention period for each of the document series (Skupsky 1994).

Many records should be kept not only because of requirements of the law, but also because the organization may need the records to defend itself in regulatory or third-party litigation or for other compliance purposes. Many companies facing environmental liability, for example, have discovered that old insurance policies and secondary evidence of insurance coverage (such as invoices, cancelled checks, and correspondence) are valuable in obtaining indemnification for past and future cleanup costs. Other companies have conducted historical records reviews to discover that predecessor and affiliated organizations may be responsible for the disposal of waste by-products. Records may be needed for a company to prove its own case as a plaintiff in a contract action and in many other types of civil litigation. A search should be made of contractual risk transfer and indemnity agreements which could limit or transfer risk.

There are other reasons to retain some records for an indefinite period of time. Each business should conduct a risk analysis to determine for itself the period of time records should be allowed to exist, as well as the location for storage. Losing valuable documents is inexcusable and destruction is irreversible.

Companies need to make careful and reasoned predictions regarding the likelihood that certain records may be useful in the future to establish a record of compliance, to defend against claims of civil or criminal liability, and to keep an accurate and complete record of business operations.

12.2.3 Development of Functional Records Retention Schedules

Once the legal basis for the retention of records is established, then a functional records retention schedule can be fully developed. There are several ways that this can be done. In most cases, businesses have to tailor the system to meet their particular needs. One possible approach is to divide the schedule into five essential steps:

- Sort the document inventory into related groups
- Develop initial retention categories, codes, and descriptions based upon the organization's needs and the scope of its business
- Revise the retention categories until each entry from the document inventory can be assigned to the appropriate retention category
- Sort the retention categories in order of the retention codes
- Assign the appropriate legal group code from the legal group file

(Skupsky 1994)

Other methods can accomplish the same objective if they are carefully planned and orchestrated by employees who have common sense, good judgment, and a decent understanding of a company's compliance requirements and objectives. Whatever system is used, however, the file information must be readily accessible—and on short notice.

12.2.4 Confidential Documents

Businesses need to be sensitive to issues involving the accessibility of confidential material. A document control system should include procedures on how to handle the creation, retention, and dissemination of confidential documents on disk or other forms of electronic communication. This will ensure that confidential documents do not get disseminated inadvertently and cause the business embarrassment or liability.

12.2.5 Integration into the Compliance System

A good document control system alone is insufficient to assure that a business will achieve regulatory compliance on all document control issues. Businesses must adopt procedures to integrate this document control system into an overall compliance system. This may begin with a one-page statement of business policy distributed to all employees regarding the reasons why accurate records, properly preserved, are an important component of business operations. For larger companies, a document control committee, made up of representatives from the legal and other departments of the business, can develop the policy and provide continuing advice regarding its interpretation (Kaplan, Murphy, and Swenson 1994).

12.2.6 Employee Training

Businesses also need to train employees on how to discard unnecessary materials. Employees need to be restrained from the creation of unnecessary documents and be encouraged to establish the business's credibility by documenting the compliance program. Employees also need guidance on how to write documents that will avoid unnecessary legal consequences in the future. If records are written in a careless fashion with hyperbole, legal conclusions by non-lawyers, speculation, and offensive language, a document control system will result in harmful evidence that can prejudice a judge or jury against the business. This is particularly true for e-mail and text messages, where employees tend to write informally and frequently do not understand the consequences of ill-chosen words and careless messages.

12.2.7 Audit

For any program to be effective, it should have a reliable audit component to identify employees who tend to save and train them regarding what should be retained and discarded. Written materials about the document control system should be distributed to each employee with specific instructions regarding the legal risk and increased costs of over-retention of documents. Over-retention may decrease computer storage capacity and add to storage and warehousing costs. Over-retention of records can add legal costs in the event of litigation without contributing to compliance efforts. An interdepartmental document control committee should have at least one member, but preferably more, who is well versed in computer technology to streamline records procedures.

12.3 The Effect of New Technologies

Because of employees' high usage of word processors, scanners, facsimile machines, electronic mail, and other technological advancements in recent years, companies must reassess their existing document control systems to determine whether new technologies are creating additional problems and how to solve these problems.

In a recent survey of large companies, an average worker received or sent 177 messages of all kinds (electronically, telephonically, or by mail) each day. The vast production of information and the time it takes to read and process information essential to business operations creates enormous problems for businesses. While computers have created many benefits for businesses, they have also caused a whole new set of problems that must be solved to make the business run smoothly. In the document control context, waste procedure improvements need to occur so that documents are not created and stored for future unintended users.

Many companies are purchasing new technologies that combine computers, facsimile machines, copiers, and printers to cut down on the unnecessary production of paper and the use of electronic energy. Service businesses like law firms, insurance companies, and financial institutions are eliminating the use of paper increasing reliance on electronically stored information. When companies are equipped with these technologies, unnecessary office space, libraries, and file rooms can be eliminated. This can cut down overhead and simplify and speed up the retrieval of information.

Most companies have solved the document-glut issue by storing information on computer "cloud" storage systems and have destroyed historical documents.

New technologies are now universally accepted for record storage purposes. Beginning on July 20, 1994, the National Archives and Records Administration (NARA) authorized electronic storage for the transfer of permanent records to the National Archives. NARA recognized that a growing number of federal agencies are using systems for the management and dissemination of federal information. Businesses have invested in such storage systems and have refining document control procedures to take advantage of these new technologies. The state-of-the-art computer technology of the 1990s is outdated in the 21st century. It will be important to monitor the development of technological advances in document control systems to make certain information stored will be able to be retrieved in the future.

12.4 Periodic Review of the System

An effective document control system is dynamic, like an environmental management system, and needs to be reviewed and revised as circumstances develop. Businesses must also be careful when amending their document control systems once they are established. Any changes to the system occurring during litigation will most likely be perceived negatively by the judge or jury. Judges have frequently imposed sanctions on lawyers and their clients for document misdeeds in the course of civil discovery.

If businesses follow these methods and obtain additional advice on how the records control procedures can be effectively integrated into an environmental management compliance system, then the records that are retained will be more easily located. Those records will then be more valuable to the business when it is in litigation or needs immediate access to respond to problems.

Environmental management systems that implement document control mechanisms effectively can improve productivity. Problems can be identified and corrected before they escalate into regulatory violations. Environmental management systems can expose inefficiencies in production and service, as well as provide information that can be useful to management to make innovations to eliminate unnecessary waste and reduce sources of pollution.

References

Kaplan, J., J. Murphy, and W. Swenson. *Compliance Programs and the Corporate Sentencing Guidelines.* New York: Clark Boardman Callaghan, 1994.

Skupsky, D.E. *Records Retention Procedures.* Denver: Information Requirements Clearinghouse, 1994.

Conclusion

Robert A. Woellner

The first two decades of the 21st century have witnessed an unprecedented change in how we value the environment. At many levels of government and throughout society, people have a greater understanding and appreciation of the need to protect our ecosystem. The only outstanding question is how much we are willing to invest, individually and collectively, in time and resources to contribute to sustainable growth for the future. As we began this book, we recognized that the assessment of environmental risk is inherently personal in nature and varies greatly between individuals, organizations, and businesses. Such issues as global warming can be divisive topics. The authors encourage the reader to treat environmental science, and, yes, global warming and COVID-19, as scientific topics, not political ones. That way we can all focus on the facts and put good minds to work solving the world's pressing issues.

Businesses must learn how to eliminate pollution, which will reduce risks and expenses and create opportunities for revenue growth. Environmental management systems may become the most expedient and effective means of transmitting the message to everyone in the organization, from the CEO and the Board of Directors to the entry-level employee, that they have an important role in any effort in minimizing environmental impacts. To take responsibility for their actions, officers, directors, and employees must understand how their individual impacts affect the environment and how they can work together in reducing future degradation. Businesses must design a meaningful message so that everyone understands that their collective performances have an extraordinary cumulative effect on the environment. One employee taking responsibility and eliminating a risk can have positive consequences and ramifications for all aspects of the business and society in general.

In their 1923 textbook, *American Problems: A Textbook in Social Progress*, which established a social studies curriculum for high school seniors, Drs. Morehouse and Graham opined that

> No fine and earnest boy or girl can study the structure of the world he lives in and know a little of its puzzling problems without wanting to do his part toward keeping what is good and curing what is bad. He wants his life to count as a positive help toward a new and better order of things. He realizes with how great a sacrifice men have bought for him the gifts that his country lays at his feet, asking only that they be well

used and passed on, undiminished and unhurt, to others to use after him. He does not wait for war or some other spectacular call, but stands ready to do his duty day by day, dependable and responsible and as well-informed as he can make himself.

What shall he do? First of all he must realize that his own individual character counts mightily. Every quality of honesty, trustworthiness, courage, and high idealism that he nurtures in himself contributes so much to the national character and to the sum of national strength. If he loves his country, he will not hurt it by allowing himself—the one citizen for whom he is definitely responsible—to be less noble than his utmost effort can make him. (Morehouse and Graham 565)

These words, published by Woellner's maternal grandmother in 1923, still resonate today.

We must not underestimate the power of collective thinking in reaching answers to the most difficult and sensitive environmental questions that pervade our society. We cannot solve these problems in a vacuum. A formal environmental management system, along with risk management and risk reduction techniques, is the most effective means of accomplishing this objective.

Reference

Morehouse, F., and S.F. Graham. *American Problems: A Textbook in Social Progress.* Boston: The Atheneum Press, Ginn and Company, 1923.

Index